Pitman Research Notes in Mathematics Series

Submission of proposals for consideration
Suggestions for publication, in the form of outlines and representative samples, are invited by the Editorial Board for assessment. Intending authors should approach one of the main editors or another member of the Editorial Board, citing the relevant AMS subject classifications. Alternatively, outlines may be sent directly to the publisher's offices. Refereeing is by members of the board and other mathematical authorities in the topic concerned, throughout the world.

Preparation of accepted manuscripts
On acceptance of a proposal, the publisher will supply full instructions for the preparation of manuscripts in a form suitable for direct photo-lithographic reproduction. Specially printed grid sheets can be provided and a contribution is offered by the publisher towards the cost of typing. Word processor output, subject to the publisher's approval, is also acceptable.

Illustrations should be prepared by the authors, ready for direct reproduction without further improvement. The use of hand-drawn symbols should be avoided wherever possible, in order to maintain maximum clarity of the text.

The publisher will be pleased to give any guidance necessary during the preparation of a typescript, and will be happy to answer any queries.

Important note
In order to avoid later retyping, intending authors are strongly urged not to begin final preparation of a typescript before receiving the publisher's guidelines. In this way it is hoped to preserve the uniform appearance of the series.

Longman Scientific & Technical
Longman House
Burnt Mill
Harlow, Essex, CM20 2JE
UK
(Telephone (0279) 426721)

Titles in this series. A full list is available from the publisher on request.

201 Riemannian geometry and holonomy groups
S Salamon
202 Strong asymptotics for extremal errors and polynomials associated with Erdös type weights
D S Lubinsky
203 Optimal control of diffusion processes
V S Borkar
204 Rings, modules and radicals
B J Gardner
205 Two-parameter eigenvalue problems in ordinary differential equations
M Faierman
206 Distributions and analytic functions
R D Carmichael and D Mitrovic
207 Semicontinuity, relaxation and integral representation in the calculus of variations
G Buttazzo
208 Recent advances in nonlinear elliptic and parabolic problems
P Bénilan, M Chipot, L Evans and M Pierre
209 Model completions, ring representations and the topology of the Pierce sheaf
A Carson
210 Retarded dynamical systems
G Stepan
211 Function spaces, differential operators and nonlinear analysis
L Paivarinta
212 Analytic function theory of one complex variable
C C Yang, Y Komatu and K Niino
213 Elements of stability of visco-elastic fluids
J Dunwoody
214 Jordan decomposition of generalized vector measures
K D Schmidt
215 A mathematical analysis of bending of plates with transverse shear deformation
C Constanda
216 Ordinary and partial differential equations. Volume II
B D Sleeman and R J Jarvis
217 Hilbert modules over function algebras
R G Douglas and V I Paulsen
218 Graph colourings
R Wilson and R Nelson
219 Hardy-type inequalities
A Kufner and B Opic
220 Nonlinear partial differential equations and their applications: Collège de France Seminar. Volume X
H Brezis and J L Lions
221 Workshop on dynamical systems
E Shiels and Z Coelho
222 Geometry and analysis in nonlinear dynamics
H W Broer and F Takens
223 Fluid dynamical aspects of combustion theory
M Onofri and A Tesei
224 Approximation of Hilbert space operators. Volume I. 2nd edition
D Herrero
225 Operator theory: proceedings of the 1988 GPOTS–Wabash conference
J B Conway and B B Morrel
226 Local cohomology and localization
J L Bueso Montero, B Torrecillas Jover and A Verschoren
227 Nonlinear waves and dissipative effects
D Fusco and A Jeffrey
228 Numerical analysis 1989
D F Griffiths and G A Watson
229 Recent developments in structured continua. Volume II
D De Kee and P Kaloni
230 Boolean methods in interpolation and approximation
F J Delvos and W Schempp
231 Further advances in twistor theory. Volume I
L J Mason and L P Hughston
232 Further advances in twistor theory. Volume II
L J Mason and L P Hughston
233 Geometry in the neighborhood of invariant manifolds of maps and flows and linearization
U Kirchgraber and K Palmer
234 Quantales and their applications
K I Rosenthal
235 Integral equations and inverse problems
V Petkov and R Lazarov
236 Pseudo-differential operators
S R Simanca
237 A functional analytic approach to statistical experiments
I M Bomze
238 Quantum mechanics, algebras and distributions
D Dubin and M Hennings
239 Hamilton flows and evolution semigroups
J Gzyl
240 Topics in controlled Markov chains
V S Borkar
241 Invariant manifold theory for hydrodynamic transition
S Sritharan
242 Lectures on the spectrum of $L^2(\Gamma\backslash G)$
F L Williams
243 Progress in variational methods in Hamiltonian systems and elliptic equations
M Girardi, M Matzeu and F Pacella
244 Optimization and nonlinear analysis
A Ioffe, M Marcus and S Reich
245 Inverse problems and imaging
G F Roach
246 Semigroup theory with applications to systems and control
N U Ahmed
247 Periodic-parabolic boundary value problems and positivity
P Hess
248 Distributions and pseudo-differential operators
S Zaidman
249 Progress in partial differential equations: the Metz surveys
M Chipot and J Saint Jean Paulin
250 Differential equations and control theory
V Barbu

251 Stability of stochastic differential equations with respect to semimartingales
X Mao
252 Fixed point theory and applications
J Baillon and M Théra
253 Nonlinear hyperbolic equations and field theory
M K V Murthy and S Spagnolo
254 Ordinary and partial differential equations. Volume III
B D Sleeman and R J Jarvis
255 Harmonic maps into homogeneous spaces
M Black
256 Boundary value and initial value problems in complex analysis: studies in complex analysis and its applications to PDEs 1
R Kühnau and W Tutschke
257 Geometric function theory and applications of complex analysis in mechanics: studies in complex analysis and its applications to PDEs 2
R Kühnau and W Tutschke
258 The development of statistics: recent contributions from China
X R Chen, K T Fang and C C Yang
259 Multiplication of distributions and applications to partial differential equations
M Oberguggenberger
260 Numerical analysis 1991
D F Griffiths and G A Watson
261 Schur's algorithm and several applications
M Bakonyi and T Constantinescu
262 Partial differential equations with complex analysis
H Begehr and A Jeffrey
263 Partial differential equations with real analysis
H Begehr and A Jeffrey
264 Solvability and bifurcations of nonlinear equations
P Drábek
265 Orientational averaging in mechanics of solids
A Lagzdins, V Tamuzs, G Teters and A Kregers
266 Progress in partial differential equations: elliptic and parabolic problems
C Bandle, J Bemelmans, M Chipot, M Grüter and J Saint Jean Paulin
267 Progress in partial differential equations: calculus of variations, applications
C Bandle, J Bemelmans, M Chipot, M Grüter and J Saint Jean Paulin
268 Stochastic partial differential equations and applications
G Da Prato and L Tubaro
269 Partial differential equations and related subjects
M Miranda
270 Operator algebras and topology
W B Arveson, A S Mishchenko, M Putinar, M A Rieffel and S Stratila
271 Operator algebras and operator theory
W B Arveson, A S Mishchenko, M Putinar, M A Rieffel and S Stratila
272 Ordinary and delay differential equations
J Wiener and J K Hale
273 Partial differential equations
J Wiener and J K Hale
274 Mathematical topics in fluid mechanics
J F Rodrigues and A Sequeira
275 Green functions for second order parabolic integro-differential problems
M G Garroni and J F Menaldi
276 Riemann waves and their applications
M W Kalinowski
277 Banach C(K)-modules and operators preserving disjointness
Y A Abramovich, E L Arenson and A K Kitover
278 Limit algebras: an introduction to subalgebras of C*-algebras
S C Power
279 Abstract evolution equations, periodic problems and applications
D Daners and P Koch Medina
280 Emerging applications in free boundary problems
J Chadam and H Rasmussen
281 Free boundary problems involving solids
J Chadam and H Rasmussen
282 Free boundary problems in fluid flow with applications
J Chadam and H Rasmussen
283 Asymptotic problems in probability theory: stochastic models and diffusions on fractals
K D Elworthy and N Ikeda
284 Asymptotic problems in probability theory: Wiener functionals and asymptotics
K D Elworthy and N Ikeda
285 Dynamical systems
R Bamon, R Labarca, J Lewowicz and J Palis
286 Models of hysteresis
A Visintin
287 Moments in probability and approximation theory
G A Anastassiou
288 Mathematical aspects of penetrative convection
B Straughan
289 Ordinary and partial differential equations. Volume IV
B D Sleeman and R J Jarvis
290 *K*-theory for real *C**-algebras
H Schröder
291 Recent developments in theoretical fluid mechanics
G P Galdi and J Necas
292 Propagation of a curved shock and nonlinear ray theory
P Prasad
293 Non-classical elastic solids
M Ciarletta and D Ieşan
294 Multigrid methods
J Bramble
295 Entropy and partial differential equations
W A Day
296 Progress in partial differential equations: the Metz surveys 2
M Chipot
297 Nonstandard methods in the calculus of variations
C Tuckey
298 Barrelledness, Baire-like- and (LF)-spaces
M Kunzinger
299 Nonlinear partial differential equations and their applications. Collège de France Seminar. Volume XI
H Brezis and J L Lions
300 Introduction to operator theory
T Yoshino

301 Generalized fractional calculus and applications
V Kiryakova
302 Nonlinear partial differential equations and their applications. Collège de France Seminar Volume XII
303 Numerical analysis 1993
D F Griffiths and G A Watson
304 Topics in abstract differential equations
S Zaidman
305 Complex analysis and its applications
C C Yang, G C Wen, K Y Li and Y M Chiang
306 Computational methods for fluid-structure interaction
J M Crolet and R Ohayon
307 Random geometrically graph directed self-similar multifractals
L Olsen
308 Progress in theoretical and computational fluid mechanics
G P Galdi, J Málek and J Necas
309 Variational methods in Lorentzian geometry
A Masiello
310 Stochastic analysis on infinite dimensional spaces
H Kunita and H-H Kuo
311 Representations of Lie groups and quantum groups
V Baldoni and M Picardello
312 Common zeros of polynomials in several variables and higher dimensional quadrature
Y Xu
313 Extending modules
N V Dung, D van Huynh, P F Smith and R Wisbauer
314 Progress in partial differential equations: the Metz surveys 3
M Chipot, J Saint Jean Paulin and I Shafrir
315 Refined large deviation limit theorems
V Vinogradov

Vladimir Vinogradov
University of Northern British Columbia, Canada

Refined large deviation limit theorems

CRC Press
Taylor & Francis Group
Boca Raton London New York

CRC Press is an imprint of the
Taylor & Francis Group, an informa business
A CHAPMAN & HALL BOOK

First published 1994 by Longman Group Limited

Published 2019 by CRC Press
Taylor & Francis Group
6000 Broken Sound Parkway NW, Suite 300
Boca Raton, FL 33487-2742

First issued in paperback 2019

ISBN 13: 978-0-367-44934-6 (pbk)
ISBN 13: 978-0-582-25499-2 (hbk)
ISSN 0269-3674

Copublished in the United States with John Wiley & Sons Inc.

AMS Subject Classifications: (Main) 60F10, 60G50, 60G70
(Subsidiary) 60E07, 60G42, 60J75

British Library Cataloguing in Publication Data

A catalogue record for this book is
available from the British Library

Library of Congress Cataloging-in-Publication Data

Vinogradov, V. (Vladimir)
Refined large deviation limit theorems / V. Vinogradov.
p. cm. -- (Pitman research notes in mathematics series ;)
Includes bibliographical references.
1. Large deviations. I. Title. II. Series.
QA273.67.V56 1994
519.2--dc20 94-33325
CIP

To my mother

Contents

Introduction 1

Chapter 1 *Asymptotic Expansions Taking into Account the Cases when the Number of Summands Comparable with the Sum is Less than or Equal to Two.* 28

1.1 Upper estimates for $|\mathbf{P}\{S_n > y\} - n \cdot c_{\alpha_1} \cdot y^{-\alpha_1}|$ 28

1.2 Asymptotic expansions of the probabilities of large deviations of S_n taking into account the case when two summands are comparable with the sum 37

1.3 Asymptotic expansions of the probabilities of large deviations of S_n in the case of quite asymmetric constraints on the asymptotic behavior of the tails 56

Chapter 2 *Asymptotic Expansions of the Probabilities of Large Deviations and Non-Uniform Estimates of Remainders in CLT.* 61

2.1 The case of power tails with integer index $\alpha_1 \geq 3$ 61

2.2 The case of power tails with index $\alpha_1 = 2$ 70

Chapter 3 *Asymptotic Expansions Taking into Account the Cases when the Number of Summands Comparable with the Sum Does not Exceed a Fixed Integer.* 75

3.1 Recursive construction of asymptotic expansions of $\mathbf{P}\{S_n > y\}$ in the case of a non-normal stable law 76

3.2 Asymptotic expansions of the probabilities of large deviations of S_n and non-uniform estimates of remainders in limit theorems on weak convergence to non-normal stable laws 105
3.3 Proof of Proposition 3.1.1 109

Chapter 4 *Limit Theorems on Large Deviations for Order Statistics.* 132
4.1 Large deviations for maxima: the tail approximation/extreme value approximation alternative 132
4.2 Large deviations for trimmed sums 141
4.3 Conditional limit theorems on large deviations for trimmed sums 162
4.4 Conditional limit theorems on weak convergence for trimmed sums 166

Chapter 5 *Large Deviations for I.I.D. Random Sums When Cramér's Condition Is Fulfilled Only on a Finite Interval.* 171
5.1 Exact asymptotics of the probabilities of large deviations and of the expectations of smooth functions of i.i.d. random sums in the case of exponential-power tails 171
5.2 Rough asymptotics of the probabilities of large deviations for i.i.d. random sums in the case when Cramér's condition is fulfilled only on a finite interval 182
5.3 Discontinuity of the most typical paths of random step-functions and the generalized concept of the action functional 186
5.4 On martingale methods for the derivation of upper estimates of the probabilities of large deviations of i.i.d. random sums 196

References 207

Foreword

This monograph is mainly devoted to studying the asymptotic behavior of the probabilities of large deviations in the case when the famous Cramér's condition is not fulfilled. In the three initial chapters, we systematically construct asymptotic expansions of the probabilities of large deviations for the classical scheme of summation of independent and identically distributed random variables. To this end, we develop the direct probabilistic method that enables one to construct expansions of increasing accuracy.

In Chapter 4, we modify this direct probabilistic method in order to investigate the asymptotics of the probabilities of large deviations of certain order statistics and their sums. Such results, in turn, provide a better understanding of the nature of formation of large deviations in the case when Cramér's condition fails.

In the conclusive Chapter 5, we treat the case when Cramér's condition is fulfilled only on a finite interval, which often takes place in various applied models. Combining a modification of the classical Cramér's techniques with some results and methods developed in the previous chapters enabled the author to establish a number of new results and also to suggest their probabilistic interpretation.

Some results of Chapter 4 are obtained in collaboration with Vladimir V. Godovan'chuk, whereas the results of the conclusive section 5.4 are joint with Don Dawson.

It should be mentioned that the preliminary versions of some results of Chapters 1, 3 and 4 comprise the author's Ph.D. thesis, written during his graduate studies in Moscow Lomonosov University under the supervision of Alexander D. Wentzell. He introduced the author to this area, which is very much appreciated.

I have had an opportunity to discuss the results of Chapter 2 with Il'dar A. Ibragimov during my visits to St. Petersburg (Leningrad) and his visit to Ottawa.

The present form of this book was made possible thanks to the generous help and support of my friends and colleagues. It has been partly assisted by NSERC Canada International Research Fellowship hosted at the Laboratory for Research in Statistics and Probability of Carleton University. Don Dawson provided his counselling and support, which brought to the publication the technical report version of the three initial chapters as well as my four abstract papers, that appeared in *Comptes Rendus - Mathematical Reports* of the Academy of Science of Canada. Robert Elliott made a number of very useful comments and, in particular, suggested the present title of this monograph. My student Yaacov Shapiro took the burden of the preparation of the camera-ready copy and did his best to make the book readable. Dmitri Verechtchiagine provided the technical assistance. During the final months of the preparation of the book, I have been supported by Don Dawson and Jon Rao.

The author expresses his gratitude to the Laboratory for Research in Statistics and Probability of Carleton University and the Department of Mathematics and Statistics of Concordia University for their very warm hospitality.

Introduction

This work is devoted to studying probabilities of large deviations for the classical scheme of summation of independent random variables $\{X_n, n \geq 1\}$ with common distribution function $F(x) := \mathbf{P}\{X_i \leq x\}$, and also to studying probabilities of large deviations for order statistics, i.e. to the investigation of asymptotics of the probabilities such as $\mathbf{P}\{X_1 + ... + X_n > y\}$ (or $\mathbf{P}\{X_1 + ... + X_n < -y\}$, $\mathbf{P}\{\max(X_1 + ... + X_n) > y\}$, where n and y vary such that these probabilities tend to zero. Note that our definition of large deviations provides a more general approach than ***the large deviation principle***, as well as the approach to large deviations as some ***refinements of theorems on weak convergence*** or ***laws of large numbers***. Set $S_n := X_1 + ... + X_n$; $S_0 := 0$.

Note that in the theory of large deviations for sums of independent random variables two polar types of the limiting behavior of the probabilities of these deviations are known. The first type is associated with the case, where the probability $\mathbf{P}\{S_n > y\}$ is generated mainly by approximately equal individual summands X_i; in this case, the asymptotics of $\mathbf{P}\{S_n > y\}$ is exponential. The fulfilment of Cramér's condition of the finiteness of the exponential moment:

$$\mathbf{E}\exp\{z \cdot X_i\} < \infty$$

for any $z \in \mathbf{R}^1$ is the typical example of this. Probabilities of large deviations under fulfilment of this condition have been extensively studied, and the results on their asymptotics up to equivalence, as well as up to rough (logarithmic) equivalence were obtained. Asymptotic expansions of $\mathbf{P}\{S_n > y\}$, valid in various ranges of deviations have also been constructed. In this respect, we mention only the works by Cramér (1938), Chernoff (1952), A.V. Nagaev (1963), Petrov (1965, 1966), Saulis and Statulevičius (1991), see also references therein.

The second polar type is characterized by the case where the main part of the probability of a large deviation $\mathbf{P}\{S_n > y\}$ is generated by one large summand X_i

comparable with the whole sum S_n. The results of the second polar type have been obtained for distribution functions F having subexponential tails such that

$$1 - F * F(x) \sim 2 \cdot (1 - F(x))$$

as $x \to \infty$, where '*' stands for the convolution. Note that the class of subexponential distributions was considered for the first time by Chistyakov (1964) in connection with certain problems of the theory of branching processes and the renewal theory. Chistyakov's work generated a great deal of interest (cf., e.g., Pinelis (1985) and references therein). Various properties of subexponential distributions were in detail studied in the monograph by Bingham, Goldie and Teugels (1987), and in many other papers.

In this work we consider distribution functions (which are typical examples of the class of subexponential distributions) satisfying the following condition:

$$(0.1) \qquad 1 - F(x) = c_{\alpha_1} \cdot x^{-\alpha_1} + o(x^{-\alpha_1})$$

as $x \to \infty$, where $\alpha_1 > 0$. In addition, we often pose an analogous restriction on the asymptotic behavior of the left-hand tail of F:

$$(0.1') \qquad F(-x) = d_{\alpha_1} \cdot x^{-\alpha_1} + o(x^{-\alpha_1}).$$

Hereinafter, we refer to the tails satisfying (0.1) or (0.1′) as **power tails.**

A wide range of the results on the exact asymptotics of the probabilities of large deviations of S_n in the case of power tails is known. Here we formulate only five of them (the most important for the sequel).

I. Let conditions (0.1) - (0.1′) be fulfilled with $\alpha_1 \in (0,1) \cup (0,2)$, and $\mathbf{E}X_i = 0$ if $\alpha_1 \in (1,2)$. Then

$$(0.2) \qquad \mathbf{P}\{S_n > y\} \sim n \cdot \mathbf{P}\{X_1 > y\} \sim n \cdot c_{\alpha_1} \cdot y^{-\alpha_1}$$

as $n \to \infty$, $y/n^{1/\alpha_1} \to \infty$. Note that relationship (0.2) was obtained in Fortus (1957), Heyde (1968) and Tkachuk (1977) under various constraints on the asymptotics of the tails of function F and on the range of variation of y; the most complete results were obtained in

Tkachuk (1977).

II. Let condition (0.1) be fulfilled with $\alpha_1 > 2$; $\mathbf{E}X_i = 0$, and $\mathbf{V}X_i = 1$. Then relationship (0.2) is valid as $n \to \infty$ with $y/\log y \geq \sqrt{n}$.

Note that this result can be found in A.V. Nagaev (1969) Theorem 1.

III. Let condition (0.1) be fulfilled with $\alpha_1 > 2$, $\mathbf{E}X_i = 0$, $\mathbf{V}X_i = 1$, and $\mathbf{E}\,|\,X_i\,|^{2+\delta} < \infty$ for some positive δ. Then

(0.3) $$\mathbf{P}\{S_n > y\} \sim 1 - \Phi(y/\sqrt{n}) + n \cdot \mathbf{P}\{X_1 > y\}$$

as $n \to \infty$, where $\Phi(\cdot)$ is the Laplace function.

Note that relationship (0.3) was obtained in A.V. Nagaev (1970). An outline of the proof can be found in S.V. Nagaev (1979) Theorem 1.9. Recently, a more subtle result of such type was obtained in Rozovskii (1993) Theorems 1-2.

IV. Let conditions (0.1) - (0.1′) be fulfilled with $\alpha_1 = 2$, and $\mathbf{E}X_i = 0$. Then for any $\varepsilon > 0$,

(0.2′) $$P\{|S_n|>y\} \sim P\{\max_{1\leq i\leq n}|X_i|>y\} \sim n \cdot P\{|X_i|>y\}$$

as $n \to \infty$ with $y \geq ((c_2 + d_2 + \varepsilon) \cdot n \cdot \log n \cdot \log\log n)^{1/2}$.

Note that relationship (0.2′) was obtained in Tkachuk (1975) Theorem 2.

Let us emphasize that the results I, II, and IV are of the second polar type of the limiting behavior of $\mathbf{P}\{S_n > y\}$, whereas the third one (relationship (0.3)) is intermediate between both polar types, as for $y \leq ((\alpha_1 - 2 - \varepsilon) \cdot n \cdot \log n)^{1/2}$ the main part of the probability of a large deviation, $\mathbf{P}\{S_n > y\}$, is generated by approximately equal individual summands X_i; for $y \geq ((\alpha_1 - 2 + \varepsilon) \cdot n \cdot \log n)^{1/2}$ the main part of this probability is generated due to the fact that one of the summands $\{X_i\}$ can be approximately as large as the whole sum S_n, and in the range of deviations $y \approx ((\alpha_1 - 2) \cdot n \cdot \log n)^{1/2}$ the parts of the probability of a large deviation generated by small individual summands and by one large summand can have the same order. Here $n \to \infty$ and ε is any fixed positive.

Note that all the results of Chapters 1 and 3 are related to the **second polar type of the limiting behavior of the probabilities of large deviations of S_n**. In Chapter 2, we consider the more complicated case of power tails with **integer** index $\alpha_1 \geq 2$. Some of the results obtained in Chapter 2 are related to the second polar type of the limiting behavior of the probabilities of large deviations of S_n, whereas, in contrast to the results of Chapters 1 and 3, subtle results (intermediate between the two polar types and analogous to (0.3)) are also given (see Corollaries 2.1.1 and 2.2.1)). In particular, the results related to integer values of index $\alpha_1 \geq 3$ (the case of normal convergence to the normal law) refine relationship (0.3). On the other hand, our results for power tails with index $\alpha_1 = 2$ (the case of non-normal convergence to the normal law) refine both relationship (0.2′), which describes the exact asymptotics of the probabilities of large deviations of S_n, and the theorem on weak convergence of the appropriately centered and normalized random sequence

$$(X_1+\ldots+X_n-n\cdot \mathbf{E}X_i) \;/\; \sqrt{(c_2+d_2)\cdot n\cdot \log n}$$

to the normal law as $n \to \infty$ (cf., e.g., Ibragimov and Linnik (1971) Theorem 2.6.5).

In Chapter 4 we review the presence of two polar types of the formation of the probabilities of large deviations from the point of view of the extreme value theory. Special consideration is given to the study of the asymptotic behavior of the maxima for typical representatives of both polar types: normal samples, and samples with power tails. The results of Chapter 4 enable us to suggest the tail approximation / extreme value approximation alternative for the case of normal samples.

In addition, we also study the asymptotics of the probabilities of large deviations of the sum S_n without one (or several) upper order statistics, in the case of power tails (see (0.1)). The results of Chapter 4 related to the values of $\alpha_1 \in (0,1) \cup (1,2)$ reveal the presence of the second polar type of the limiting behavior of the probabilities of large deviations. Namely, the main part of the probability of a large deviation of S_n - $\max(X_1, \ldots, X_n)$ is generated mainly by two individual summands X_i and X_j, which can be comparable with the whole sum S_n. In contrast, some of the results of Sections 4.2 - 4.4 of Chapter 4, related to the values of $\alpha_1 > 2$ are in fact subtle and intermediate between the polar types. Thus, it is shown that in the range of large deviations $y \approx (g(\alpha_1) \cdot n \cdot \log n)^{1/2}$, the parts of the probability of a large deviation generated by small individual summands and by one large summand can have the same order (These results are analogous to relationship (0.3).) Here, $g(\alpha_1)$ is a certain function of α_1. In the same chapter, we also establish the result of a more complicated nature (related to the case of power tails with index $\alpha_1 > 2$), which describes the asymptotic behavior of the conditional probabilities of large deviations of S_n - $\max(X_1, \ldots, X_n)$, in the range of large deviations of S_n.

The conclusive Chapter 5 is devoted to the investigation of the probabilities of large deviations of the sum S_n without upper restrictions on the range of values of those deviations, in the case where the Cramér's condition of finiteness of exponential moments is fulfilled only on a finite interval. In this illustrative chapter, we use some of the results and the methods developed in Chapters 1 - 4.

Now, let us proceed with a more detailed description of our results. The first chapter is devoted to studying the probabilities of large deviations of S_n in the case of power tails with index $\alpha_1 \in (0,1) \cup (1,2) \cup (2,\infty)$, both for $n \to \infty$ and for n being fixed. Note that in the latter case, the following result can be viewed as an analog to (0.2).

V. Let condition (0.1) be fulfilled with α_1 being any positive. Then relationship (0.2) is valid for any fixed integer $n \geq 1$ as $y \to \infty$ (cf. Feller (1971) Vol. II, Chapter 8, (8.14)).

Note that for the theory of the limit theorems, the case, where $n \to \infty$ is more interesting, but we cannot restrict ourselves to this case only, since in order to construct more precise asymptotic expansions for $\mathbf{P}\{S_n > y\}$, related to the case $n \to \infty$, we will need less precise asymptotic expansions for $\mathbf{P}\{S_n > y\}$, and for $\mathbf{P}\{S_n < -y\}$ in the case of fixed n. Hence, we have to consider both cases. We prefer to simultaneously demonstrate the results and their proofs to be suitable for both cases.

In Section 1.1, we refine relationship (0.2). Obviously, (0.2) itself implies only that the expression

$$(0.4) \qquad |\mathbf{P}\{S_n > y\} - n \cdot c_{\alpha_1} \cdot y^{-\alpha_1}|$$

is $o(n \cdot y^{-\alpha_1})$, so for a not sufficiently large n, it cannot be too small. More precise upper estimates for expression (0.4) are easily derived from Theorem 1.1.1 proved in Section 1.1. Let us emphasize that the estimate of Theorem 1.1.1 is too general to provide exact rates of convergence in (0.4) for all considered values of α_1. However, Theorem 1.1.1 yields various corollaries (related to different values of index α_1), which in their turn can be used in the construction of the asymptotic expansions for $\mathbf{P}\{S_n > y\}$ and $\mathbf{P}\{S_n < -y\}$. Such Corollaries 1.1.1 - 1.1.3 can also be found in Section 1.1. Note that Corollary 1.1.2 (related to the case of normal convergence to the normal law) is conceptually close to the results of Kim and A.V. Nagaev (1972), where the upper estimates for

$$|\mathbf{P}\{S_n > y\}/(n \cdot \mathbf{P}\{X_1 > y\}) - 1|$$

were obtained.

Sections 1.2 and 1.3 are devoted to the construction of the asymptotic expansions for $\mathbf{P}\{S_n > y\}$. It is carried out under fulfilment of a stronger condition (compare to (0.1)). Namely, we require that the right-hand tail of $F(x)$ admits expansion over negative powers of x as $x \to \infty$:

$$1 - F(x) = \sum_{i=1}^{\ell} c_{\alpha_i} \cdot x^{-\alpha_i} + o(x^{-r}), \tag{0.5}$$

where $\alpha_1 > 0$ is any real; $\alpha_1 < \alpha_2 < \ldots < \alpha_\ell \leq r$. Let us emphasize that in Ibragimov and Linnik (1971) there was suggested an investigation of the asymptotics of the probabilities of large deviations of the sum of independent random variables with common density $\mathbf{p}(x)$ having the following asymptotics as $x \to \infty$:

$$\mathbf{p}(x) = \int_a^{Ka} d\mathrm{G}(\nu)/x^{\nu} + o(x^{-Ka-\varepsilon}), \tag{0.6}$$

where $a > 0$, $K > 1$, $\varepsilon > 0$, and function $\mathrm{G}(\cdot)$ has bounded variation. For a special case when $\mathrm{G}(\cdot)$ is a pure jump function with a finite number of jumps, the 'local' condition (0.6) is analogous to our 'integral' condition (0.5).

In Sections 1.2 - 1.3, we also assume additional constraints on the asymptotics of the left-hand tail of $F(x)$; the most restrictive of them is condition (0.1′). Accuracy of asymptotic expansions for $\mathbf{P}\{S_n > y\}$ constructed in these sections depends on the assumed constraints.

In Chapter 2 we separately consider the cases of power tails with integer values of index $\alpha_1 \geq 3$ (cf. Section 2.1) and with index $\alpha_1 = 2$ (cf. Section 2.2). The reason for the latter case to be separated is its complicated nature, namely, this case is related to the phenomenon of **non-normal convergence to the normal law.** However, the method of deriving the results of the both sections of Chapter 2 is the same. First, the exact (in the sense of dependence on n and on y) non-uniform estimates of the remainder in CLT are established under fulfilment of slightly stronger conditions, compare to (0.1) - (0.1′) (cf. Theorems 2.1.1 and 2.2.1). Thereupon, by analogy with the proof of Theorem 1.2.2 of Section 1.2, we apply the results of Theorems 2.1.1 and 2.2.1 for the derivation of the asymptotic expansions of the probabilities of large deviations of S_n, under fulfilment of Condition (0.5).

Let us point out that though Theorems 2.1.1 and 2.2.1 play only the auxiliary role in this work, they have their own value, because the results of these theorems can be viewed as an extension (Theorem 2.1.1) and a generalization (Theorem 2.2.1) of the Edgeworth expansions in CLT. It should be also noted that the concept of '**pseudomoments in Ibragimov's sense**' turns out to be useful in Chapter 2 (a simplified concept of such pseudomoments was considered by Ibragimov (1967)). Let us emphasize that our Theorem 2.1.1 refines Theorems 1 - 2 by Ibragimov (1967) and the theorem by Michel (1976). In addition, note that the results of Chapter 2 are based on the author's works, Vinogradov (1990a) devoted to the case of integer $\alpha_1 \geq 3$, and Vinogradov (1992c) devoted to the case of $\alpha_1 = 2$.

It should be noted that in fact, Chapter 2 demonstrates that in the most complicated cases only an interplay between the direct probabilistic method and the method of characteristic functions can lead to some progress in deriving the asymptotics of the probabilities of large deviations of S_n.

In Chapter 3, we consider only the case $\alpha_1 \in (0,1) \cup (1,2)$. Significantly more precise asymptotic expansions for $\mathbf{P}\{S_n > y\}$ (compare to those from Chapters 1 and 2) are constructed recursively along with analogous expansions for $\mathbf{P}\{S_n < -y\}$ (see Section 3.1). The initial expansion for $\mathbf{P}\{S_n > y\}$ is comprised of only one term, $n \cdot c_{\alpha_1} \cdot y^{-\alpha_1}$; it is, in fact, a restatement of Corollary 1.1.1 from Section 1.1. In order to construct such expansions, the fulfilment of both condition (0.5) and its left-hand analog is required:

$$(0.5') \qquad F(-x) = \sum_{i=1}^{\ell} d_{\alpha_i} \cdot x^{-\alpha_i} + o(x^{-r})$$

as $x \to \infty$.

The main result of Section 3.1 is Theorem 3.1.1. Its proof is based on a series of auxiliary results; the most complicated of those is Proposition 3.1.1, which is only formulated in Section 3.1. The proof of Proposition 3.1.1 is cumbersome and thus deferred to Section 3.3. In Section 3.2, the example illustrating Theorem 3.1.1 in a special case of non-negative random variables satisfying Condition (0.5) with $\alpha_1 \in (1/2, 2/3) \cup (2/3, 1)$; $\ell = 2$, $\alpha_2 = 2 \cdot \alpha_1$, $r = \alpha_1 + 1$ and

$$c_{\alpha_2} = -c_{\alpha_1}^{\,2} \cdot \Gamma(1-\alpha_1)^2 \,/\, (2 \cdot \Gamma(1-2\alpha_1))$$

is considered in detail.

The present work mainly employs a simple, direct probabilistic method. Let us briefly describe its basic features (separate points of the description will be repeated in a more detailed way in the corresponding parts of Chapters 1, 2, 3, and 4). First, the probability $\mathbf{P}\{S_n > y\}$ is written as

$$(0.7) \qquad \mathbf{P}\{S_n > y\} = \mathbf{P}\{S_n > y,\, X_1 \le \theta y, \dots, X_n \le \theta y\} + \mathbf{P}\{\bigcup_{i=1}^{n} \{S_n > y, X_i > \theta y\}\}.$$

Here $\theta > 0$ is assumed to be any real; in the sequel θ will be chosen to be fixed (not depending on n and y), but sufficiently small: the more terms of the asymptotic expansion of $\mathbf{P}\{S_n > y\}$ under consideration, the smaller should be the value of θ chosen.

The first probability on the right-hand side of (0.7) is relatively easy to estimate (for a sufficiently small θ), and the second probability can be rewritten (by the use of the well-known formula for the probability of union of non-disjoint events) as

$$(0.8) \qquad \sum_{m=1}^{n} (-1)^{m+1} \cdot \binom{n}{m} \mathbf{P}\{S_n > y,\, X_{n-m+1} > \theta y, \dots, X_n > \theta y\}.$$

We drop the terms of this alternating sum beginning with a certain number (the second in Sections 1.1, 1.3 and Chapter 2; the third in Section 1.2; the term with an arbitrary fixed number in Section 3.1); the absolute value of the emerged error is known to be bounded by the absolute value of the first omitted term. The mth term of the sum (0.8) coincides (up to a certain multiplier) with

$$\mathbf{P}\{S_{n-m} + X_{n-m+1} + \ldots + X_n > y,\ X_{n-m+1} > \theta y,\ \ldots,\ X_n > \theta y\},$$

and this probability can be expressed as follows:

(0.9) $$\mathbf{P}\{S_{n-m} + \ldots + X_n > y,\ X_{n-m+1} > \theta y,\ \ldots,\ X_n > \theta y\} =$$

$$\mathbf{P}\{X_1 > \theta y\}^m - \int_{-\infty}^{(1-m\theta)y} \Pi_m(y,z)\cdot dF_{S_{n-m}}(z)\ ,$$

where $\Pi_m(y,z) := \mathbf{P}\{X_1 > \theta y,\ \ldots,\ X_m > \theta y,\ S_m \leq y - z\}$. Note that in order to obtain (0.9), the integration of the tails of the distribution functions $F_{S_{n-m}}$ should be used. It is not difficult to demonstrate that functions $\{\Pi_m\}$ can also be defined by recursion:

$$\Pi_0(y,z) \equiv 1;$$

$$\Pi_{m+1}(y,z) = \int_{\theta y}^{(1-m\theta)y} \Pi_m(y,z+v)\cdot dF(v) \quad \text{if } z \leq (1-(m+1)\cdot\theta)\cdot y;$$

$$\Pi_{m+1}(y,z) = 0 \quad \text{otherwise.}$$

Note that under fulfilment of (0.5), the right-hand tail of function F is approximated by a finite linear combination of power functions as $x \to +\infty$. Similarly, we now introduce (by recursion) functions $H_m(y,z)$ which depend on a finite number of parameters and are smooth in z, as follows:

$$H_0(y,z) \equiv 1;$$

$$H_{m+1}(y,z) = \int_{\theta y}^{(1-m\theta)y} H_m(y,z+v)\cdot d\Big(-\sum_{i=1}^{l} c_{\alpha_i}\cdot v^{-\alpha_i}\Big) \quad \text{if } z \leq (1-(m+1)\cdot\theta)\cdot y;$$

$$H_{m+1}(y,z) = 0 \quad \text{otherwise.}$$

Asymptotics of functions H_m is easily derived. Note that if only the first term of the sum (0.8) is considered (Sections 1.1, 1.3 and Chapter 2) we do not need to deal with functions H_m at all; if the initial two terms of the sum (0.8) are considered (Section 1.2) then functions $H_1(y,z)$ and $H_2(y,z)$ (presented in certain integrals) should be replaced by

$$\sum_{s=0}^{[\alpha_1]} \frac{\partial^s H_1(y,z)}{\partial z^s}\Big|_{z=0} \cdot \frac{z^s}{s!}$$

and by H_2 $(y,0)$, respectively. If the initial k terms of the sum (0.8) are considered (Section 3.1) then functions H_m (y,z) (for $1 \le m \le k$) should be replaced by

$$\sum_{s=0}^{[(k-m)\alpha_1]} \frac{\partial^s H_m(y,z)}{\partial z^s}\Big|_{z=0} \cdot \frac{z^s}{s!}, \quad \text{etc.}$$

Recall that the asymptotic expansion for $\mathbf{P}\{S_n > y\}$, taking into account the initial k+1 terms of the sum (0.8), is constructed recursively by the use of the antecedent expansions which take into account the first, the initial two, ... , the initial k terms of the sum (0.8). To this end, the probability presented in the mth term of the sum (0.8) is rewritten by means of formula (0.9), function Π_m (as it was already noted) may be replaced by

$$\sum_{s=0}^{[(k+1-m)\alpha_1]} \frac{\partial^s H_m(y,z)}{\partial z^s}\Big|_{z=0} \cdot \frac{z^s}{s!},$$

and the distribution function $F_{S_{n-m}}(z)$ is approximated by the use of the antecedent expansions for $\mathbf{P}\{S_{n-m} > z\}$ if $z > 0$ and for $\mathbf{P}\{S_{n-m} < z\}$ if $z < 0$, which take into account the initial k+1-m terms of the sum (0.8). Recall that these expansions constructed in previous steps of recursion are valid under symmetric constraints on the right-hand and left-hand tails of $F(x)$ (Conditions (0.5)-(0.5′)). These expansions are denoted in Section 3.1 by

$$\sum_{j:\beta_j \le (k+1-m)\alpha_1} c_{\beta_j}(n-m)\cdot z^{-\beta_j} \quad \text{and} \quad \sum_{j:\beta_j \le (k+1-m)\alpha_1} d_{\beta_j}(n-m)\cdot z^{-\beta_j}.$$

Integrals obtained after these changes are easily evaluated up to required accuracy.

In order to estimate the error terms which emerge due to such changes, the concept of the **pseudomoments in Cramér's** sense turns out to be useful. Note that these pseudomoments were introduced in the author's work, Vinogradov (1985b). The definition and some properties of the pseudomoments in Cramér's sense can be found in Section 3.1, Lemmas 3.1.1-3.1.4. If Conditions (0.5)-(0.5′) are fulfilled with all $\{\alpha_i, 1 \le i \le \ell\}$ being non-integers then the following integrals would be finite for any non-negative integer

$k \leq [r]$:

$$\int_{-\infty}^{0} x^{k} \cdot d\,(F(x) - \sum_{\substack{1 \leq t: \\ \alpha_t \leq k}} d_{\alpha_t} \cdot |x|^{-\alpha_t})$$

and

$$\int_{0}^{+\infty} x^{k} \cdot d\,(F(x) + \sum_{\substack{1 \leq t: \\ \alpha_t \leq k}} c_{\alpha_t} \cdot |x|^{-\alpha_t}).$$

Let us denote the sum of these integrals by $b_k\,(X_i\,)$. Hereinafter, we refer to $b_k\,(X_i\,)$ as the kth pseudomoment in Cramér's sense. It turns out that under a certain supplementary constraint on the tail behavior of function F (cf. condition (3.1.6) in Section 3.1), the sum S_n also possesses finite pseudomoments $b_k\,(S_n\,)$ in Cramér's sense up to $k \leq [r]$ such that $b_1(S_n),\dots,b_k(S_n)$ are expressed in terms of $b_1(X_n),\dots,b_k(X_n)$ by the same formulas as the standard moments.

Hence, the terms of the sum (0.8) from the first to the kth, can be evaluated up to required accuracy. The latter $(k+1)^{st}$ term, which has to be taken into account in this step is estimated as easily as constructing the initial expansion for $\mathbf{P}\{S_n > y\}$, in which only the first term of the sum (0.8) was taken into account. Let us note that in a special case $n \to \infty$, and $y/n^{1/\alpha_1} \to \infty$, the asymptotics of the kth term, which is equal to

$$(-1)^{k+2} \binom{n}{k+1} \cdot \mathbf{P}\{S_n > y, X_{n-k} > \theta y,\dots,X_n > \theta y\}$$

can be derived from Theorems 1 - 2 by Godovan'chuk (1978). To this end, we rewrite the latter expression as

$(-1)^{k+2} \cdot \mathbf{P}\{S_n > y,\ \exists$ integer $1 \leq i_1 < \dots < i_{k+1} \leq n$ such that $X_{i_1} > \theta y,\dots,X_{i_{k+1}} > \theta y\}$

(recall that $\alpha_1 \in (0,1) \cup (0,2)$). The latter probability in turn, coincides with the probability of the event that the random step-function (r.s.f.)

$$S_{n,y}\,(t) := S_{[nt]}\,/y$$

considered in the interval [0,1] belongs to the set A_{k+1} of the càdlàg space $\mathbf{D}[0,1]$ equipped with the uniform metric ρ, where

$A_{k+1} := \{f \in D[0,1]$ satisfying the following conditions: $\exists$ time instants $0 < t_1 < \dots < t_{k+1} \leq 1$ such that $\min_{1 \leq j \leq k+1} (f(t_j) - f(t_j -)) > \theta,\ f(1) > 1\}$.

Applying Theorems 1 - 2 by Godovan'chuk (1978) for the derivation of the asymptotics of events $\mathbf{P}\{S_{n,y} \in A_{k+1}\}$ we get that

$$\binom{n}{k+1} \cdot \mathbf{P}\{S_n > y, X_{n-k} > \theta y, \dots, X_n > \theta y\}$$

$$\sim \; c_{\alpha_1}^{k+1} \cdot \frac{(ny^{-\alpha_1})^{k+1}}{(k+1)!} \cdot \int\limits_{\substack{u_1 + \dots + u_{k+1}: \\ \min(u_1, \dots, u_{k+1}) > \theta}} \dots \int d(-u_1^{-\alpha_1}) \dots d(-u_{k+1}^{-\alpha_1}) \, .$$

Therefore, the terms of the sum (0.8) from the first to the $(k+1)^{st}$ can be evaluated up to the required accuracy, and we obtain the asymptotic expansion for $\mathbf{P}\{S_n > y\}$, which takes into account the $k+1$ initial terms of the sum (0.8). This expansion is used in the sequel when constructing the asymptotic expansion for $\mathbf{P}\{S_n > y\}$, which takes into account the $k+2$ initial terms of the sum (0.8), etc.

Let us note that the asymptotic expansion for $\mathbf{P}\{S_n > y\}$, which is related to the case of $\alpha_1 \in (0,1) \cup (0,2)$, and takes into account only the two initial terms of the sum (0.8) (cf. Theorem 1.2.1 of Section 1.2) coincides with the expansion of Theorem 3.1.1 of Section 3.1 for k=2, but the first one is proved under weaker constraints on the asymptotics of the left-hand tail of $F(x)$ (compare (0.1′) to (0.5′)). Let us emphasize that when constructing all the expansions for $\mathbf{P}\{S_n > y\}$ in the case of $\alpha_1 \in (0,1) \cup (0,2)$, only the direct probabilistic method described above is used. In contrast, when constructing asymptotic expansions for $\mathbf{P}\{S_n > y\}$ in the case of $\alpha_1 > 2$ (cf. Theorem 1.2.2 of Section 1.2 and Theorems 1.3.1′, 1.3.2 of Section 1.3) we cannot restrict ourselves to the use of this direct probabilistic method alone due to a more complicated nature of the formation of large deviations in this case (compare (0.3) to (0.2)). Namely, one should apply non-uniform estimates of the remainder in CLT, which can be derived from Theorem 4 of Bikjalis (1966) in the case, where $2 < \alpha_1 < 3$; and from Theorem 1 of Osipov (1972) in the case of $\alpha_1 > 3$. Note that the asymptotic expansionfor $\mathbf{P}\{S_n > y\}$ of Section 1.2, related to the case of non-integer $\alpha_1 > 3$, is not valid in the full range of large deviations, but valid only for $y \geq Const \cdot n^{1/2+\kappa}$, where $\kappa > 0$. In this range of deviations we have

$$1-\Phi(y/n^{1/2}) = o(n\cdot y^{-\alpha_1})$$

and hence, by (0.3),

$$\mathbf{P}\{S_n > y\} \sim n\cdot c_{\alpha_1}\cdot y^{-\alpha_1}.$$

It is interesting to note that for the special case of fixed values of the parameter n, the results of Theorems 1.2.1 and 1.3.1, are conceptually close to Theorem 3 of Geluk and Pakes (1991), related to a certain subclass $\mathcal{L}^2$ of subexponential distributions, and to Theorem 1.2(i′) and (ii′) of Geluk (1992). However, note that distributions with regularly varying tails (and hence with power tails) belonging to the class of subexponential distributions, generally do not belong to the subclass $\mathcal{L}^2$ (see the remark to Corollary 2 in Geluk and Pakes (1991) for discussion).

In Section 1.3, we construct asymptotic expansions for $\mathbf{P}\{S_n > y\}$ under fulfilment of condition (0.5) and weaker restrictions on the left-hand tail behavior of function F (compare to Section 1.2); these results generalize the results of Section 1.2 for a wider class of distributions. The main results of Section 1.3 are Theorems 1.3.1, 1.3.1′, and 1.3.2.

In Section 1.3 along with (0.5) we require the fulfilment of the following condition:

There exists a positive constant K such that

(0.10) **for some $\beta > 0$ and for any positive x**

$$F(-x) + 1 - F(x) \le K\cdot x^{-\beta}.$$

Theorems 1.3.1 and 1.3.1′ are related to the cases of $\beta \in (0,1) \cup (1,2)$, and non-integer $\beta > 2$, respectively.

Another result of this section, Theorem 1.3.2, is related to the case of $\alpha_1 > 2$ (not necessarily non-integer). Along with fulfilment of (0.5) it is also assumed that $\mathbf{E}|X_1|^\beta < \infty$ for some $2 \le \beta < \alpha_1$. Note that similarly to the proof of Theorem 1.2.2, along with the direct probabilistic method described above, the non-uniform estimates of the remainder in CLT, which can be derived from Corollary 1 and Theorem 4 of Bikjalis (1966) in the case where $2 \le \beta \le 3$; and from Theorem 1 of Osipov (1972) in the case where $\beta > 3$ have to be used. In addition, in Section 1.3 we consider a simple Example 1.3.1, illustrating

Theorem 1.3.2 in the case, where condition (0.5) is fulfilled with the remainder, $o(x^{-r})$, being equal to zero identically, and r.v.'s $\{X_i\}$ are only assumed to have a finite variance. It should be noted that Example 1.3.1 implies the exactness (in the sense of dependence on n and on y) of an upper estimate for expression (0.4) obtained by Kim and A.V. Nagaev (1972).

As we already mentioned, a different approach to the derivation of the asymptotics of the probabilities of large deviations of S_n consists in their derivation from limit theorems on **normal deviations** (i.e. on weak convergence) with non-uniform estimates of the remainder (Integral and local theorems of such kind can be found, e.g., in Bentkus and Bloznelis (1989) and in Inzhevitov (1986), respectively.) In Example 3.2.1 of Section 3.2, we compare the results which can be derived from our Theorem 3.1.1 (proved by the direct probabilistic method) and from the non-uniform estimate in the limit theorem on weak convergence to a non-normal stable law obtained by Bentkus and Bloznelis (1989) (proved by the method of convolutions).

Note that the main results of Chapters 1 and 3 were announced in Vinogradov (1992a); very special cases of some results of Chapter 1 were published in the following author's works, Vinogradov (1984, 1985a). Besides, a special case of Theorem 1.2.1 (obtained by the author of the present monograph) was included in the monograph by Wentzell (1990) Theorem 6.3.2.

In Chapter 4, we investigate the asymptotic behavior of the *probabilities of large deviations for order statistics and their sums*, i.e., the asymptotics of the probabilities such as $\mathbf{P}\{\min(X_1,...,X_n) > y\}$, as well as $\mathbf{P}\{S_n - \max(X_1,...,X_n) > y\}$, and $\mathbf{P}\{\max(X_1,...,X_n) > y\}$, etc., where n and y vary such that these probabilities tend to zero. In Section 4.1, we reveal the presence of two polar types of the formation of large deviations (rare events) and also review this fact from the point of view of the extreme value theory. Special consideration is given to the study of the asymptotic behavior of maxima for typical representatives of both polar types: normal samples and samples with the tails of the power type. The tail approximation/extreme value approximation alternative is suggested for the case of the normal sample. Note that by the *extreme value approximation* we mean the approximation of the probabilities of large deviations of $\max(X_1,...,X_n)$ in terms of the tails of limiting distributions for these maxima, whereas by the *tail approximation* we mean the approximation of the probabilities of large deviations of $\max(X_1,...,X_n)$ by means of a variant of the direct probabilistic method described above (see also formulas (0.12) and (0.12′) below).

It should be mentioned that historically, the standard normal sample often played the role of a 'guinea pig' for the extreme value theory. Thus, it is well known that in this case, the distribution of the properly centered and normalized maximum $X_1(n)$ converges weakly to the Gumbel (double exponential) distribution $\Lambda(x) := \exp\{-e^{-x}\}$:

(0.11) $$P\{\max(X_1,...,X_n) \leq A_n + B_n \cdot x\} \Rightarrow \Lambda(x)$$

as $n \to \infty$, where

$$A_n := (2 \cdot \log n)^{1/2} - 1/2(\log\log n + \log(4\pi)) \cdot (2 \cdot \log n)^{-1/2},$$

and $$B_n := (2 \cdot \log n)^{-1/2}.$$

Note that (0.11) remains true if A_n and B_n are replaced by a_n and b_n respectively, such that $b_n/B_n \to 1$ and $(a_n - A_n)/B_n \to 0$ as $n \to \infty$ (see, e.g., Gnedenko (1943) Lemma 2.2.2). In addition, it was known since the works by Tippett (1925); and by Fisher and Tippett (1928) (cf. also Gumbel (1958)) that the rate of convergence in (0.11) is very slow and worst on the tails (in contrast to the classical CLT, where the remainder equals zero identically in the case of the normal samples). Let us now quote the following rigorous result. Hall (1979) obtained that the exact bounds in (0.11) are as follows:

There exists a positive constant C such that for any integer $n \geq 2$,

(0.11′) $$\frac{C}{\log n} \leq \sup_{x \in \mathbb{R}^1} |P\{X_1(n) \leq A'_n + B'_n \cdot x\} - \Lambda(x)| \leq \frac{3}{\log n},$$

where $$B'_n := (A'_n)^{-1},$$

and $$A'_n := (2 \cdot \log n)^{-1/2} - (\log\log n + \log(4\pi)) \cdot (8 \cdot \log n)^{-1/2}$$
$$- [(\log\log n + \log(4\pi))^2 - 4(\log\log n + \log(4\pi)] \cdot (8 \cdot (2 \cdot \log n)^{3/2})^{-1}.$$

In addition, in order to emphasize the fact that the rate of convergence in (0.11) is very slow and worst on the tails, we now quote the following remark given on p. 492 of Hall (1980): "... *an approximation by an extreme value distribution is of little use in determining a critical point for the rejection of outliers. What is needed is a non-uniform estimate* $Q_n(x)$ of $\mathbf{P}\{\max(X_1,...,X_n) \le x\}$. *Ideally such an estimate should be simple to calculate, and the relative error*

$$|Q_n(x) - \mathbf{P}\{\max(X_1,...,X_n) \le x\}| \, / \, (1 - \mathbf{P}\{\max(X_1,...,X_n) \le x\})$$

should tend to zero as x and n tend to infinity."

Hall (1980) succeeded in determining one non-uniform estimate $Q_n(x)$ of such kind, but his method seems to be applicable to the case of normal samples only. However, it turns out that one can develop an alternative method (hereinafter referred to as the **tail approximation**), which provides asymptotic expansions for the probabilities of the right-hand large deviations along with accurate estimates of remainders. The just mentioned method is similar to the direct probabilistic method reviewed above from the point of view of the classical scheme of summation of i.i.d.r.v.'s, and is based on the following apparent representation, which is equally applicable to the case of an arbitrary distribution function F of i.i.d. random variables $\{X_n, n \ge 1\}$ (We do not even require the distribution function of properly centered and normalized maxima $\max(X_1,...,X_n)$ to belong to the domain of attraction of any limiting distribution):

$$\begin{aligned} \mathbf{P}\{\max(X_1,...,X_n) > y\} &= \sum_{k=1}^{n} (-1)^{k+1} \cdot \mathbf{P}\{\exists i_1,...,i_k : X_{i_1} > y,...,X_{i_k} > y\} \\ &= \sum_{k=1}^{n} (-1)^{k+1} \cdot \binom{n}{k} \cdot \mathbf{P}\{X_i > y\}^k. \end{aligned} \tag{0.12}$$

It is obvious that (0.12) follows from the apparent representation that

$$\mathbf{P}\{\max(X_1,...,X_n) > y\} = \mathbf{P}\Big(\bigcup_{i=1}^{n} \{X_i > y\}\Big) \tag{0.12$'$}$$

and the fact that the latter probability can be rewritten by means of the well-known formula for the probability of union of non-disjoint events.

It should be mentioned that this approach is basically analogous to the *direct probabilistic method* described above and developed in Chapters 1-3, but much simpler. In this respect, note that formulas (0.12) - (0.12′) are analogous to formulas (0.7) and

(0.8).

In the same section, we also study the asymptotics of the probabilities of large deviations of $\max(X_1,...,X_n)$ from the sample of i.i.d. random variables whose common distribution function satisfies condition (0.1). Note that it was proved in Gnedenko (1943) Theorem 4 that under fulfilment of (0.1), the distribution function of the properly normalized $\max(X_1,...,X_n)$

$$(0.13) \qquad F_n(x) := \mathbf{P}\{\max(X_1,...,X_n) \le (c_{\alpha_1} \cdot n)^{1/\alpha_1} \cdot x\} \to \Psi_{\alpha_1}(x)$$

as $n \to \infty$, where the limiting distribution $\Psi_{\alpha_1}(\cdot)$ is defined as

$$\Psi_{\alpha_1}(x) := \exp\{-x^{-\alpha_1}\} \quad \text{for } x > 0, \text{ and}$$
$$\Psi_{\alpha_1}(x) := 0 \qquad \text{otherwise}.$$

It is also well known that under fulfilment of (0.1), the following result on the asymptotic behavior of the probabilities of large deviations is valid:

$$(0.14) \qquad \mathbf{P}\{\max(X_1,...,X_n) > y\} \sim n \cdot c_{\alpha_1} \cdot y^{-\alpha_1} \sim 1 - \Psi_{\alpha_1}\left(\left(\frac{y}{c_{\alpha_1} \cdot n}\right)^{-\alpha_1}\right)$$

as $n \to \infty$, $y/n^{1/\alpha_1} \to \infty$ (cf., e.g., Resnick (1987) Section 2.3).

Theorem 4.1.2 of Section 4.1 provides an asymptotic expansion for the distribution of $\max(X_1,...,X_n)$ refining both relationship (0.13) on weak convergence to $\Psi_{\alpha_1}(\cdot)$ and relationship (0.14) on the exact asymptotics of the probabilities of large deviations. The results of Section 4.1 were announced in Vinogradov (1992b) and Vinogradov (1994). To illustrate the results of Section 4.1, we compare them with certain results obtained by de Haan and Hordijk (1972), Hall (1980), and Smith (1982).

In Sections 4.2 and 4.3, we obtain the results on large deviations of the sum S_n without one (or several) upper order statistics (hereinafter, we refer to such quantities as *trimmed sums*). Note that although the results of those sections do not follow from the results of Chapters 1 - 3, they are conceptually close to each other. In order to make the connections clear, let us formulate the following 'conditioned' result of the second polar type (which is due to Durrett (1980) Theorem 3.1):

Let condition (0.1) be fulfilled with $\alpha_1 > 2$, $\mathbf{E}X_i = 0$, and $\mathbf{V}X_i < \infty$. Then for any fixed

$u > 0$,

$$(0.15) \qquad \{S_{n,n}(\cdot) \mid S_{n,n}(1) > u\} \to J_{u,\alpha_1} \cdot I_{\{U \le (\cdot)\}}$$

as $n \to \infty$, where I_A is the indicator of the set A, and r.v. J_{u,α_1}, which represents the magnitude of the maximal jump of r.s.f. $S_{n,n}(\cdot)$, has the following distribution:

$$\mathbf{P}\{J_{u,\alpha_1} > x\} = (x/u)^{-\alpha_1} \quad \text{for } x \ge u,$$

and r.v. U which represents the time instant of the maximal jump of r.s.f. $S_{n,n}(\cdot)$ is uniformly distributed on the unit interval [0,1].

Note that in view of (0.15), when studying some problems related to the investigation of the asymptotic behavior of the sum of a sample of growing size, it would be natural to replace it by $\max(X_1,\ldots,X_n)$ (or more generally, by the sum of a finite number of upper order statistics). Hence, we need results that could be applicable for the estimation of errors emerged due to such replacements. For other reasons regarding importance of the problem of studying the asymptotic behavior of sums after the removal of extreme terms (i.e., trimmed sums) we refer to Maller (1982), Introduction. At the same time, in the work by Maller (1982) it was considered to be more important, in view of possible statistical decisions, to remove the summands with maximal absolute value. The asymptotics of the probabilities of large deviations of such sums are given in Theorems 4.2.4 - 4.2.5 of Section 4.2. The results on weak convergence of such (properly centered and normalized) trimmed sums were established in Arov and Bobrov (1960), Hall (1978), and Teugels (1981) (see also Csörgő, Hauesler and Mason (1991) for a comprehensive list of references).

Recall that two polar types of the limiting behavior of the probabilities of large deviations also emerge for certain sums of order statistics. The results related to the first polar (or Cramér's) type were obtained, e.g., in Bentkus and Zitikis (1990), and Zitikis (1991); whereas the results of the second polar type were established in Gafurov and Khamdamov (1987). In contrast, some results of Sections 4.2 - 4.3 of this work are intermediate between the polar types. The results of these sections supplement the studies on the unconditional asymptotics of distributions of trimmed sums in the case of power tails. In this respect, let us mention that the result on weak convergence of such sums was obtained in Csörgő, Csörgő, Horváth and Mason (1986) Theorem 4.1, and the exact rate of convergence in LLN was derived in Hatori, Maejima and Mori (1980) Theorem 2.

In Section 4.2, we prove a series of subtle results on the exact asymptotics of the probabilities of large deviations for trimmed sums under fulfilment of Condition (0.1) and certain restrictions on the behavior of $F(x)$ on the left-hand semi-axis (we assume the fulfilment of (0.1′) or (0.10) with $\beta = \alpha_l$). The main method of the derivation of the results of this section consists in transferring the problem to the càdlàg space $\mathbf{D}[0,1]$. The results of this section describing the asymptotic behavior of the probabilities of large deviations of trimmed sums can be interesting from the point of view of possible applications. On the other hand, they also provide a better understanding of the nature of large deviations. To explain this, we now consider the realizations of r.s.f. $S_{n,y}(\cdot)$ in $\mathbf{D}[0,1]$ equipped with the uniform metric ρ. The question arises is what the typical paths of $S_{n,y}(\cdot)$ are like if an event of small probability (a large deviation) has occurred (i.e., r.s.f. $S_{n,y}$ hits a set separated from the function identically equal to zero, in the uniform metric). It is well known that if the r.s.f. $S_{n,y}(\cdot)$ is constructed starting from random variables with finite exponential moments, then a large deviation is mainly contributed by close-to-continuous or even close-to-smooth paths (cf., e.g., Wentzell (1990) Introduction) - the result of the first polar type. On the other hand, under fulfilment of condition (0.1) with $\alpha_1 < 2$, large deviations of $S_{n,y}(\cdot)$ occur mainly via almost piecewise constant paths, which perform one or several big jumps - the result of the second polar type. Note that this fact was established in Godovan'chuk (1978) Theorems 1-2 for the full range of large deviations and in the case of $\alpha_1 < 2$.

However, for the case of fulfilment of condition (0.1) with index $\alpha_1 > 2$ (note that in this case the weak convergence of $S_{n,\sqrt{n}}(\cdot)$ to the Wiener process $\mathbf{w}(\cdot)$ holds) the paths of both types can give comparable contributions to the probability of a large deviation - the result intermediate between the both polar types. Let us emphasize that this is possible for the case of $\alpha_1 > 2$ and in a very narrow range of large deviations only (cf., e.g., Pinelis (1981) for the case in which $S_{n,y}$ performs at most one big jump). Outside this range of deviations, if $y \geq n^{1/2+\chi}$, events of small probability occur mainly due to the almost piecewise constant paths (here $\chi > 0$ being any real). This fact follows from Theorems 1′, 2′ and Remark 4 of Godovan'chuk (1981).

Therefore, it seems reasonable to assume that large deviations of r.s.f. $S_{n,y}$ in D[0,1] may sometimes occur via the paths which perform one or several big jumps and close-to-continuous functions between them.

The truthfullness of this hypothesis is established in Theorems 4.2.1, 4.2.2, and 4.2.4.ii of Section 4.2 (see also Remarks 4.2.1 and 4.2.4 therein).

The methods employed in Section 4.2 are essentially purely probabilistic ones. They involve the splitting of probabilities of events that r.s.f. $S_{n,y}$ hits certain sets of D[0,1] into several parts. The latter ones are estimated by the use of various techniques similar to those developed in Chapter 1 of the present monograph, as well as in Godovan'chuk (1978, 1981) and in Pinelis (1981).

It should be mentioned that preliminary versions of the results of Sections 4.2 - 4.3 have been obtained during author's collaborative research with V.V. Godovan'chuk (see Vinogradov and Godovan'chuk (1989); and Godovan'chuk and Vinogradov (1990)).

In Section 4.4, we establish the conditional results on weak convergence of trimmed sums in the range of large deviations of S_n or $\max(X_1,...,X_n)$ (see Theorems 4.4.1 and 4.4.2 therein). Let us emphasize that conditional weak limits for trimmed sums emerged in those theorems differ from unconditional ones established in Csörgő, Csörgő, Horváth and Mason (1986). Note that preliminary versions of points (i) of these theorems obtained by the author of the present monograph were announced in Vinogradov and Godovan'chuk (1989) Theorems 5 - 6.

In Chapter 5, we derive the exact asymptotics of the probabilities of large deviations and of the densities of sums S_n of i.i.d.r.v.'s as well as of the expectations of smooth functions of S_n / n in the case when the Cramér's condition is fulfilled only on a finite interval. However, the probabilistic interpretation of certain results of Chapter 5 differs from that relevant to the famous Cramér's theorem (see Theorem 5.3.1 of Section 5.3.)

In Sections 5.1 and 5.3, we consider the sequence $\{X_n, n \geq 1\}$ of i.i.d.r.v.'s with common distribution function $F(\cdot)$, having the following asymptotics on the right-hand tail:

(0.16) $$1 - F(x) \sim C_a \cdot \exp\{-z_0 \cdot x\} \cdot x^{-a}$$

as $x \to \infty$, where $z_0 > 0$, $C_a > 0$, and $a > 2$. Hereinafter, we refer to such tails as **exponential-power tails**.

Let us point out that distributions satisfying (0.16) are closely related to the class of the generalized inverse Gaussian distributions that has been thoroughly studied in the recent past as from the point of view of various applications, as well as due to the fact that distributions belonging to this class exhibit a large number of interesting properties (cf., e.g., Jørgensen (1982), Chhikara and Folks (1989) and references therein).

It is obvious that (0.16) implies that Cramér's condition of finiteness of exponential moments is fulfilled only on a finite interval, namely,

$$\Psi(z) := \mathbf{E}\exp\{z \cdot X_i\} < \infty \qquad \text{for } z \in [0, z_0\,],$$

whereas

$$\Psi(z) = \infty \qquad \text{for } z > z_0\,.$$

Let us point out that we do not make any assumption on the behavior of $F(\cdot)$ on the left-hand tail.

In Section 5.1 we reveal that under fulfilment of modifications of condition (0.16), the exact asymptotics of the probabilities of large deviations of S_n, as well as of the densities $p_{S_n}(\cdot)$ of S_n are not given in terms of Cramér's formula. Naturally, it is true only in a certain range of large deviations (see Theorems 5.1.1 - 5.1.2 therein). These results are related to the fact that under fulfilment of (0.16), the rate function

$$\mathbf{H}(u) := \sup_z\, [zu - V(z)],$$

is not strongly downward convex (here, $V(z) := \log \Psi(z)$).

In Section 5.3, we present the probabilistic interpretation of this phenomenon by establishing conditioned limit theorems for r.s.f. $\eta_n(t) := S_{[nt]} / n$, where $t \in [0,1]$ and $S_0 := 0$ (see Theorem 5.3.1 therein). In the same section, we also establish that this phenomenon is in fact linked with the generalized concept of the **action functional**, relevant to a wide class of possibly discontinuous functions (see Proposition 5.3.1 therein).

Note that the just mentioned generalized concept of the action functional was first introduced by Lynch and Sethuraman (1987) in the context of families of processes with independent increments (treated in their work as probability measures on [0,1]) and then developed by Mogulskii (1993) in the context of families of processes with independent increments taking values in the space **D**[0,1], equipped with the Skorokhod or the uniform topology.

In addition, in Section 5.1 we also derive the exact asymptotics of expectations of $F(S_n / n)$ for a certain class of smooth functions F (see Theorem 5.1.3). Note that this result is of the same spirit as the results obtained by Dubrovskii (1976) (see also their version in Chapter 5 of the monograph by Wentzell (1990)). Let us point out that some versions of the results of Sections 5.1 and 5.3 were presented in Vinogradov (1983, 1990b, 1993).

Note that one is often interested in results on the asymptotic behavior of the probabilities of large deviations only up to logarithmic equivalence (such results are sometimes called **rough** limit theorems on large deviations). In this respect, we present one result of such type in Section 5.2 (see Theorem 5.2.1 therein). In this theorem, we establish that for two very general classes of distributions, for which the Cramér's condition of the finiteness of exponential moments is fulfilled only on finite intervals,

$$\log \mathbf{P}\{S_n > y\} \sim - n \cdot \mathrm{H}(y/n) \tag{0.17}$$

as $n \to \infty$ in a very wide range of large deviations of y. Note that Theorem 5.2.1 refines some results obtained in Vinogradov (1985c).

It should be mentioned that in the works on large deviations of the first (or Cramér's) type, relationships of the same kind as (0.17) are often being established. The results similar to (0.17) proved to be very useful as for establishing a large number of interesting results of the theory of Stochastic Processes and Differential Equations (cf., e.g., Freidlin and Wentzell (1984)), as well as for derivation of various applications (cf., e.g., Borovkov and Mogulskii (1992, 1993)). Moreover, such results are closely related to the **large deviation principle**, which is one of the corner stones of the modern theory of large deviations and is, in fact, associated with a large deviation counterpart of the Prokhorov theorem (cf., e.g., Puhalskii (1991)). However, the reason that we have devoted the entire Section 5.2 to the proof of (0.17) is not just to follow the tradition, but rather to use (0.17) for the proof of Proposition 5.3.1 of Section 5.3.

Let us point out that in Point (i) of Theorem 5.2.1 of Section 5.2, we establish relationship (0.17) as $n \to \infty$ with $y \geq n \cdot (\mathbf{E}X_i + \varepsilon)$, for a class of distributions that contains distributions satisfying (0.16). On the other hand, the exact asymptotics of $\mathbf{P}\{S_n > y\}$ in this range of deviations is given by two different formulas (see (5.1.6) and (5.1.12)) having different probabilistic interpretation. However, Theorem 5.2.1.i demonstrates that as long as we are concerned with the asymptotics of $\mathbf{P}\{S_n > y\}$ only up to logarithmic equivalence, the two different exact asymptotics are glued together.

Recall that our consideration of distributions satisfying (0.16) is motivated by various problems, where such distributions emerge. In particular, distributions having exponential moments only on finite intervals or not strongly downward convex rate functions (functionals) arise in the classical theory of branching processes, as well as in the modern theory of measure-valued processes. They also arise in various models of the Actuarial Science (cf., e.g., Klüppelberg (1989)). Thus, Chover, Ney and Wainger (1973) considered the subcritical age-dependent process whose particle lifetime distribution $G(\cdot)$ belongs to a certain class $C(d)$ (which is closely related to the class of distributions satisfying (0.16)). Here,

$$\boldsymbol{d} := \max \left(\rho: \int_0^\infty e^{\rho \cdot t} \cdot dG(t) < \infty \right).$$

Let m denote the average number of children born at each birth epoch. In particular, Chover, Ney and Wainger (1973) established that if a subcritical branching process starts with one particle at the initial moment, $G(\cdot) \in \{C(d), d > 1\}$ and $\boldsymbol{d} \cdot m < 1$ (this means that the Malthusian parameter does not exist) then the total number of descendants of the initial particle alive at time t, conditioned on non-extinction, converges to a non-degenerate limit as $t \to \infty$ (see Theorem 4 therein). However, the generation number converges to a finite limit as $t \to \infty$ in a certain sense, which suggests that the class $\{C(d), d > 1\}$ plays a borderline role for branching processes (cf. Section 5 of that paper for more details). Moreover, it was established in the recent work by Fleischmann and Kaj (1992) that for a certain family of measure-valued processes, the rate functional (which plays the role analogous to the rate function H(·) and is related to the action functional) is not strongly downward convex and even not continuous as a rule (see Theorem 5.2.1 and Example 5.2.4 therein).

In the conclusive Section 5.4 of the present monograph, we also consider random variables whose distributions satisfy the Cramér's condition of the finiteness of exponential moments only on finite intervals. In fact, the consideration of those random variables was necessitated by the recent developments in the theory of measure-valued processes. In particular, Theorem 5.4.1 of Section 5.4 can be used for establishing the mutual singularity of genealogical structures of two important families of measure-valued processes (see Dawson and Vinogradov (1994) for details).

Now, let us give some auxiliary notation. Let $z(u)$ denote the unique root to the equation $V'(z) = u$. Let $u_0 := \sup\{u\text{: such that } z(u) \text{ exists}\}$. Then it is easily seen that under fulfilment of (0.16),

(0.18) $$u_0 = V'(z_0\text{-}) < \infty.$$

Recall that under fulfilment of (0.16), the rate function H(·) is not strongly downward convex. Namely,

$$\mathrm{H}(u) = uz(u) - V(z(u)) \qquad \text{for } u \leq u_0\,,$$

and

$$\mathrm{H}(u) = \mathrm{H}(u_0) + z_0(u - u_0) \qquad \text{for } u > u_0.$$

Now, let us briefly describe the main tools used in Sections 5.1 - 5.3.

First, we introduce the family of the probability measures $\mathbf{P}^z$ (with $z \in [0,z_0]$) on the space of infinite sequences

$$\Omega := \{\omega = (x_1,\ldots,x_n,\ldots)\text{: } X_i(\omega) = x_i\}$$

as follows: Let μ^0 be the probability measure on $\mathbb{R}^1$ generated by the distribution function F, and define the new probability measure μ^z on $\mathbb{R}^1$ by means of the following formula:

$$\mu^z\,(du) := e^{zu} \cdot \mu^0\,(du) \,/\, \Psi(z).$$

Set

$$\mathbf{P}^z := \mu^z \otimes ... \otimes \mu^z \otimes ...$$

It is easily seen that although the probability measures $\mathbf{P}^0$ and $\mathbf{P}^z$ are mutually singular, they are absolutely continuous on the σ-algebras generated by any finite set of $(X_1,...,X_n)$ such that the Radon-Nykodim derivatives are as follows:

$$(0.19) \qquad \frac{d\mathbf{P}^z}{d\mathbf{P}^0}\Big|_{\sigma(X_1,...,X_n)} = \frac{\exp\{z(X_1+....+X_n)\}}{\Psi(z)^n}$$

and

$$(0.19') \qquad \frac{d\mathbf{P}^0}{d\mathbf{P}^z}\Big|_{\sigma(X_1,...,X_n)} = \frac{\exp\{-z(X_1+....+X_n)\}}{\Psi(z)^{-n}}$$

Let $\mathbf{P} := \mathbf{P}^0$, and denote the projections of the probability measures $\mathbf{P}$ and $\mathbf{P}^z$ to the σ-algebras $\sigma(X_1,...,X_n)$ by $\mathbf{P}_n(\cdot)$ and $\mathbf{P}_n^z(\cdot)$, respectively. Then (0.19) can be rewritten in the integral form for each $A \in \sigma(X_1,...,X_n)$ as follows:

$$(0.19'') \qquad \mathbf{P}^z_n(A) = \Psi(z)^{-n} \cdot \mathbf{E}^z_n(I_A \exp\{z_0 \cdot (X_1 +...+ X_n)\}),$$

where $\mathbf{E}^z_n$ is the expectation with respect to the probability measure $\mathbf{P}^z_n$, and I_A is the indicator of the set A.

Note that the above transformation of measures goes back to Cramér (1938) and is often called Cramér's transformation. Let us point out that the expectation and the variance of r.v.'s X_n with respect to $\mathbf{P}_n^{z(u)}$ are as follows: $\mathbf{E}_{z(u)}X_i = V'(z(u)) = u$, and $V_{z(u)}X_i = V''(z(u))$. In particular, it can be shown that under fulfilment of (0.16), $\mathbf{E}_z X_i = u_0$, and $V_z X_i = V''(z_0 -)$.

Now, note that under fulfilment of (0.16), the exact asymptotics of

$$\mathbf{P}\{S_n > nu\}$$

as $n \to \infty$ for $u \in (\mathbf{E}X_i, u_0]$ is expressed in terms of the famous Cramér's formula (cf., e.g., Cramér (1938)). The idea of the proof consists in employing the Cramér's transformation of measures (see (0.19) and (0.19′)) with the subsequent use of the various refinements of CLT for studying the asymptotic behavior of

$$\mathbf{P}_n^{z(u)}\{S_n > nu\} = \mathbf{P}_n^{z(u)}\{S_n - \mathbf{E}_{z(u)}S_n > 0\}.$$

In Section 5.1, we derive the exact asymptotics of $\mathbf{P}\{S_n > y\}$ as $n \to \infty$, $y - nu_0 \to \infty$.

However, in contrast to the Cramér's case, we cannot reduce our problem to the consideration of the probabilities of **normal deviations** of S_n with respect to any transformed measure $\mathbf{P}_n^z$. This follows from (0.16), since we cannot make $\mathbf{E}_z X_i$ greater than u_0 (< y/n). However, we can still consider the distribution of S_n with respect to the 'rightmost' measure $\mathbf{P}_n^{z_0}$ reducing our problem to studying the probabilities of **large deviations** of S_n :

$$(0.20) \qquad \mathbf{P}_n^{z_0}\{S_n - \mathbf{E}_{z_0} S_n > y - nu_0\}$$

as $n \to \infty$, $y - n \cdot u_0 \to \infty$.

It turns out that under fulfilment of (0.16), the asymptotics of the right-hand tail of the distribution of X_i with respect to $\mathbf{P}_1^{z_0}$ is of **power type**. This fact enables one to use the results of Chapters 1 - 3 of this monograph related to the asymptotic behavior of the **probabilities of large deviations** of S_n in the **case of power tails** in order to derive the exact asymptotics of (0.20). This is clarified by a purely analytical Lemma 5.1.1 of Section 5.1. In particular, in Lemma 5.1.1 we establish that under fulfilment of (0.16), the common distribution function $\mathbf{F}^{z_0}(\cdot)$ of the random sequence $\{X_n, n > 1\}$ satisfies the following property, with respect to the transformed measure $\mathbf{P}_1^{z_0}$:

$$(0.21) \qquad |F^{z_0}(y + x) - F^{z_0}(y)| \sim \frac{z_0 \cdot c_a}{\Psi(z_0)} \cdot |x| \cdot y^{-a}$$

as $y \to \infty$, where $0 < Const \leq |x| = o(y)$. Note that Condition (0.21) was introduced by A.V. Nagaev (1969) formula (6).

In turn, the fact that Condition (0.21) is fulfilled makes possible an application of **interval limit theorems** on large deviations in the case of power tails obtained by A.V. Nagaev (1969) Theorem 3 and by Tkachuk (1973) Theorem 1, with respect to the transformed measure $\mathbf{P}_n^{z_0}$. In particular, this enables one to derive the asymptotics of (0.20), and the asymptotics of smooth functions of S_n / n.

It was already mentioned that the conclusive Section 5.4 was motivated by the theory of measure-valued processes. On the other hand, in order to derive the upper bounds for the probabilities of large deviations for random sequences comprised of linear combinations of infinite numbers of exponential (1) variables, given in this section, we use the martingale approach which was only recently developed in Liptser and Shiryayev (1989); and Liptser and Pukhalskii (1992). This makes clear one of our goals - to demonstrate the power of martingale methods for relatively simple random sequences, for which one can explicitly compute stochastic exponentials. The reason that we have included the results derived by an application of martingale methods to the chapter in which a quite different technique is mainly employed is due to the fact that as in Sections 5.1 - 5.3, as well as in Section 5.4, we consider the distributions that satisfy the Cramér's condition of the finiteness of exponential moments only on finite intervals. In addition, it seems that some results of the present monograph can be naturally generalized for various families of martingales or supermartingales. In particular, it is already known that under fulfilment of a modification of Cramér's condition, the limiting behavior of the probabilities of large deviations of martingales and semimartingales is of the first polar type (cf., e.g., Liptser and Pukhalskii (1992)). Moreover, at least one result of the second polar type for martingales is also known (cf. Fuk (1973)).

To complete the introduction to the monograph, we provide some notation that has not yet been introduced, and will be used below. In this work, we use the triple numeration of formulas: the first, second, and third digits denote numbers of a chapter, a section in the chapter, and a formula in the section, respectively. Analogous formulas can be enumerated with the same digits but subsequent formulas possess extra prime (or primes) to the latter digit. Theorems, propositions, lemmas, etc. are enumerated according to the same rule. However, figures are enumerated by one digit throughout the monograph. In the Introduction we use the double numeration of formulas: the first digit is 0, and the second one is the number of a formula. Sections are enumerated with two digits: the first one is the number of a chapter, and the second one is the number of a section in the chapter. Different positive constants, which can depend on $\{\alpha_i\}$, $\{c_{\alpha i}\}$, $\{d_{\alpha i}\}$ (but do not depend neither on n nor on y) are denoted by K or C. Different positive constants presented in the same formula are denoted by $K_1,\ldots,K_m$ or $C_1,\ldots,C_m$; $a \vee b$ and $a \wedge b$ stand for $\max(a,b)$ and $\min(a,b)$, respectively. End of proof is marked by $\square$.

Chapter 1

Asymptotic Expansions Taking into Account the Cases when the Number of Summands Comparable with the Sum is Less than or Equal to Two.

1.1 Upper Estimates for $|\mathbf{P}\{S_n > y\} - n\cdot c_{\alpha_1}\cdot y^{-\alpha_1}|$.

In this section, we refine some results by Fortus (1957), Heyde (1968), A.V.Nagaev (1969), Feller (1971), Vol. II, Chapter 8, (8.14); and Tkachuk (1975, 1977), where the representation

$$\mathbf{P}\{S_n > y\} \sim n\cdot c_{\alpha_1}\cdot y^{-\alpha_1}$$

was established in various ranges of variation of the parameters n and y, under fulfilment of Condition (0.1) and other various constraints on the left-hand tail behavior of F.

First, we prove the following theorem which is valid for an arbitrary sequence of independent identically distributed random variables $\{X_n, n \geq 1\}$, without any constraint on their common distribution function $F(\cdot)$.

Theorem 1.1.1. Let $\alpha_1 > 0$, $c_{\alpha_1} > 0$, and $0 < \kappa < 2/3$ be any fixed real numbers. Let $\phi(\cdot,\cdot)$ be some function from $\mathbf{N} \times \mathbf{R}_+^1$ into $\mathbf{R}_+^1$, such that for any integer $n \geq 1$ and for any real $y > 0$, we have $0 \leq \phi(n,y) \leq \kappa y$. Then for any integer $n \geq 1$ and for any real $y > 0$,

$$|\mathbf{P}\{S_n > y\} - n\cdot c_{\alpha_1}\cdot y^{-\alpha_1}| \leq \mathbf{P}\{\max_{1\leq k\leq n} S_k > y/3\}^2 + \binom{n}{2}\cdot\mathbf{P}\{X_1 > y/3\}^2$$

$$(1.1.1) \qquad + n\cdot c_{\alpha_1}\cdot x^{-\alpha_1}\cdot [\ \mathbf{P}\{S_{n-1} \leq -2y/3\} + (3^{\alpha_1}-1)\cdot(\mathbf{P}\{S_{n-1} > 2y/3\}$$

$$+ \mathbf{P}\{\phi(n,y) \leq |S_{n-1}| \leq 2y/3\}) + \alpha_1\cdot(1-\kappa)^{-\alpha_1-1}\cdot\phi(n,y)/y\ ]$$

$$+ n\cdot\sup_{x\geq y/3}|1 - F(x) - c_{\alpha_1}\cdot x^{-\alpha_1}|.$$

It should be mentioned that estimate (1.1.1) can have its own value, though we will need only its corollaries refining some of the just quoted results by Fortus (1957), Heyde (1968), A.V.Nagaev (1969), Feller (1971), and Tkachuk (1975, 1977). Let us note that the rightmost term on the right-hand side of (1.1.1) tends to zero as $y \to \infty$ only under fulfilment of Condition (0.1). Otherwise, the above inequality is less informative.

Proof of Theorem 1.1.1. We prefer to prove the theorem by the unified method for any integer $n \geq 1$, such that both, the formulation and the proof are suitable as in the case $n \to \infty$, as well as for fixed n.

It is easily seen that for any positive y and u,

$$\mathbf{P}\{S_n > y\} = \mathbf{P}\{S_n > y, \max(X_1,\dots,X_n) \leq u\} + \mathbf{P}\Big(\bigcup_{l=1}^{n} \{S_n > y, X_l > u\}\Big).$$

Applying the known formula for the probability of a union of non-disjoint events to the second term we get that

$$\mathbf{P}\{S_n > y\} = \mathbf{P}\{S_n > y, \max(X_1,\dots,X_n) \leq u\}$$

$$(1.1.2) \qquad + \sum_{m=1}^{n} (-1)^{m+1} \cdot \binom{n}{m} \cdot \mathbf{P}\{S_n > y, X_{n-m+1} > u, \dots, X_n > u\}.$$

Now, we proceed with two lemmas valid for an arbitrary sequence of i.i.d.r.v.'s $\{X_n, n \geq 1\}$. These lemmas will be used as in the proof of Theorem 1.1.1, as well as for the derivation of further refinements of the asymptotics of $\mathbf{P}\{S_n > y\}$. These lemmas, 1.1.1 and 1.1.2 are related to the leftmost and the rightmost terms on the right-hand side of (1.1.2), respectively.

Lemma 1.1.1. For any integer $n \geq 1$ and $\ell \geq 1$, and for any real $y > 0$ with $u \leq y/(2\cdot\ell-1)$,

$$(1.1.3) \qquad \mathbf{P}\{\max_{1\leq k\leq n} S_k \geq y, \max(X_1,\dots,X_n) \leq u\} \leq \mathbf{P}\{\max_{1\leq k\leq n} S_k \geq u\}^{\ell}.$$

Proof of Lemma 1.1.1. Note that the idea of the proof of this lemma is similar to that of Lemma 2 of Godovan'chuk (1978). First, assume that $n \geq 2\cdot\ell - 1$ and prove inequality

(1.1.3) by induction on ℓ. The induction base is trivial; let us prove the induction step. To this end, we assume the validity of (1.1.3) for $\ell = m$ and then derive it for $\ell = m + 1$. For this purpose, we introduce the stopping time τ_u by analogy with Godovan'chuk (1978):

$$\tau_u := \min(k \le n:\ X_1 + \dots + X_k \ge u), \quad \text{if such } k \text{ exists};$$
$$\tau_u := n \quad \text{otherwise}.$$

Recall that here $n \ge 2\cdot m + 1$ is a fixed integer. It is easily seen that

$$\mathbf{P}\{\max_{1\le k\le n} S_k \ge y,\ \max(X_1,\dots,X_n) \le u\} \tag{1.1.4}$$
$$= \sum_{s=1}^{n-1} \mathbf{P}\{\max_{1\le k\le n} S_k \ge y,\ \tau_u = s,\ \max(X_1,\dots,X_n) \le u\}$$

since $u \le y/(2\cdot m+1)$. The right-hand side of (1.1.4) does not exceed the following sum:

$$\sum_{s=1}^{n-1} \mathbf{P}\{\tau_u = s\}\cdot \mathbf{P}\{\max_{s+1\le j\le n}(S_j - S_s) \ge y-2u,\ \max(X_1,\dots,X_n) \le u\} \tag{1.1.5}$$
$$\le \mathbf{P}\{\max_{1\le k\le n} S_k \ge u\}\cdot \mathbf{P}\{\max_{1\le k\le n-2} S_k \ge y-2u,\ \max(X_1,\dots,X_{n-2}) \le u\}.$$

Note that the inequalities $u \le y/(2m+1)$ and $n-2 \ge 2m-1$ imply that $u \le (y-2u)/(2m-1)$. Hence, we are able to apply the induction hypothesis in order to estimate the second factor on the right-hand side of (1.1.5):

$$\mathbf{P}\{\max_{1\le k\le n-2} S_k \ge y-2u, \max(X_1,\dots,X_{n-2}) \le u\} \le \mathbf{P}\{\max_{1\le k\le n-2} S_k \ge u\}^m \le \mathbf{P}\{\max_{1\le k\le n} S_k \ge u\}^m.$$

A combination of these inequalities with (1.1.4) and (1.1.5) yields the induction step, and the assertion of the lemma is proved for $n \ge 2\ell - 1$.

Now, assume that $n < 2\ell - 1$. Then a combination of the condition $\max(X_1, \dots, X_n) \le u$ with the inequality $u \le y/(2\ell - 1)$ yields that for any integer $1 \le k \le n$,

$$S_k \le k\cdot u \le n\cdot y/(2\ell - 1) < y.$$

Therefore,

$$\mathbf{P}\{\max_{1\le k\le n} S_k \ge y,\ \max(X_1,\dots,X_n) \le u\} = 0. \quad \square$$

Lemma 1.1.2. If in the sum on the right-hand side of (1.1.2) only the $k - 1$ initial terms are retained, while all the subsequent terms are dropped, then the error (i.e., the true value minus the approximation) acquires the sign of the first omitted term, namely, $(-1)^k$, and is smaller in absolute value, i.e., for any integer $1 \le k \le n$,

$$(1.1.6)\qquad \begin{aligned} &\Big|\sum_{m=k}^{n} (-1)^{m+1}\cdot\binom{n}{m}\cdot \mathbf{P}\{S_n > y,\ X_{n-m+1} > u,\dots,X_n > u\}\Big| \\ &\le \binom{n}{k}\cdot \mathbf{P}\{S_n > y,\ X_{n-k+1} > u,\dots,X_n > u\}. \end{aligned}$$

Proof of Lemma 1.1.2 consists in application of Bonferroni's Inequalities (cf., e.g., Feller (1971), Vol. I, Chapter IV, (5.7). □

Now, let us proceed with the proof of Theorem 1.1.1. Note that for $n = 1$, inequality (1.1.1) is obviously true. Thus, we can assume that $n \ge 2$. Applying equality (1.1.2) with $u = y/3$, Lemma 1.1.2 with $k = 2$, and $u = y/3$ we obtain that

$$\begin{aligned} &\big|\mathbf{P}\{S_n>y\}-n\cdot \mathbf{P}\{S_n>y, X_n>y/3\}\big| \le \mathbf{P}\{S_n>y, \max(X_1,\dots,X_n)\le y/3\} \\ &+ \binom{n}{2}\cdot \mathbf{P}\{S_n > y,\ X_{n-1}> y/3,\ X_n> y/3\}. \end{aligned}$$

To estimate the first term on the right-hand side of this inequality, we apply Lemma 1.1.1 with $\ell = 2$, and $u = y/3$; in order to estimate the second term we get rid of the condition $\{S_n > y\}$ and use the fact that $\{X_n, n \ge 1\}$ are i.i.d.r.v.'s. Hence,

$$(1.1.7)\qquad \begin{aligned} &|\mathbf{P}\{S_n > y\} - n\cdot \mathbf{P}\{S_n > y,\ X_n > y/3\}| \\ &\le \mathbf{P}\{\max_{1\le k\le n} S_k \ge y/3\}^2 + \binom{n}{2}\cdot \mathbf{P}\{X_1 > y/3\}^2. \end{aligned}$$

Therefore, in order to complete the proof of the theorem, it only remains to estimate

$$|\mathbf{P}\{S_n > y,\ X_n > y/3\} - c_{\alpha_1}\cdot y^{-\alpha_1}|.$$

It is easily seen that

$$\mathbf{P}\{S_n > y,\ X_n > y/3\} = \mathbf{P}\{S_{n-1} > 2y/3\} \cdot \mathbf{P}\{X_n > y/3\}$$

$$(1.1.8) \qquad + \int_{-\infty}^{2y/3} \mathbf{P}\{X_n > y-z\} \cdot dF_{S_{n-1}}(z) := \Pi_1 + \Pi_2,$$

where

$$\Pi_1 := c_{\alpha_1} \cdot (y/3)^{-\alpha_1} \cdot \mathbf{P}\{S_{n-1} > 2y/3\}$$

$$(1.1.9) \qquad + \int_{-\infty}^{2y/3} c_{\alpha_1} \cdot (y-z)^{-\alpha_1} \cdot dF_{S_{n-1}}(z),$$

and

$$\Pi_2 := (\ \mathbf{P}\{X_n > y/3\} - c_{\alpha_1} \cdot (y/3)^{-\alpha_1}) \cdot \mathbf{P}\{S_{n-1} > 2y/3\}$$

$$(1.1.10) \qquad + \int_{-\infty}^{2y/3} (\mathbf{P}\{X_n > y - z\} - c_{\alpha_1} \cdot (y-z)^{-\alpha_1}) \cdot dF_{S_{n-1}}(z).$$

We first estimate $|\Pi_2|$. Obviously,

$$|\Pi_2| \le \sup_{x \ge y/3} |1 - F(x) - c_{\alpha_1} \cdot x^{-\alpha_1}| \cdot (\mathbf{P}\{S_{n-1} > 2y/3\} + \int_{-\infty}^{2y/3} dF_{S_{n-1}}(z))$$

$$(1.1.11) \qquad = \sup_{x \ge y/3} |1 - F(x) - c_{\alpha_1} \cdot x^{-\alpha_1}|.$$

Note (running ahead) that the latter expression multiplied by n is an upper estimate for $n \cdot |\Pi_2|$.

The asymptotics of $\mathbf{P}\{S_n > y\}$ is mainly contributed by $n \cdot \Pi_1$, hence we proceed with studying this expression. It follows from (1.1.9) that

$$\Pi_1 - c_{\alpha_1} \cdot y^{-\alpha_1} = c_{\alpha_1} \cdot ((y/3)^{-\alpha_1} - y^{-\alpha_1}) \cdot \mathbf{P}\{S_{n-1} > 2y/3\}$$

$$c_{\alpha_1} \cdot \int_{-\infty}^{2y/3} ((y-z)^{-\alpha_1} - y^{-\alpha_1}) \cdot dF_{S_{n-1}}(z).$$

We break up the latter integral into the integrals from $-\infty$ to $-2y/3$, and from $-2y/3$ to $2y/3$. It is easily seen that the absolute value of the integrand in the first integral does not exceed $y^{-\alpha_1}$. Hence,

$$\text{(1.1.12)}\qquad \begin{aligned} |\Pi_1 - c_{\alpha_1} \cdot y^{-\alpha_1}| &\le c_{\alpha_1} \cdot (3^{\alpha_1} - 1) \cdot y^{-\alpha_1} \cdot \mathbf{P}\{S_{n-1} > 2y/3\} \\ &+ c_{\alpha_1} \cdot y^{-\alpha_1} \cdot \mathbf{P}\{S_{n-1} \le -2y/3\} + c_{\alpha_1} \cdot y^{-\alpha_1} \cdot \int_{-2y/3}^{2y/3} |(1 - z/y)^{-\alpha_1} - 1| \cdot dF_{S_{n-1}}(z). \end{aligned}$$

To estimate the latter integral, we also break it up into the integrals over $\phi(n,y) < |z| \le 2y/3$ and over $|z| \le \phi(n,y)$. The first of them does not exceed

$$\text{(1.1.13)}\qquad \begin{aligned} &\max\{|(1 - z/y)^{-\alpha_1} - 1| : \phi(n,y) < |z| \le 2y/3\} \cdot \mathbf{P}\{\phi(n,y) < |S_{n-1}| \le 2y/3\} \\ &= (3^{\alpha_1} - 1) \cdot \mathbf{P}\{\phi(n,y) < |S_{n-1}| \le 2y/3\}, \end{aligned}$$

and the second one does not exceed

$$\max\{|(1 - z/y)^{-\alpha_1} - 1| : |z| \le \phi(n,y)\}.$$

Note that the function $(1 - z/y)^{-\alpha_1}$ is monotonic and convex, which implies that this maximum is equal to

$$\text{(1.1.14)}\qquad (1 - \phi(n,y)/y)^{-\alpha_1} - 1 \le \alpha_1 \cdot (1 - \kappa)^{-\alpha_1 - 1} \cdot \phi(n,y)/y.$$

Combining (1.1.7) - (1.1.14) we obtain the assertion of Theorem 1.1.1. □

Now, let us demonstrate the range of application of Theorem 1.1.1. For this purpose, we apply this theorem to derive more precise results on the asymptotics of $\mathbf{P}\{S_n > y\}$ than those obtained in the works by Fortus (1957), Heyde (1968), A.V.Nagaev (1969), Feller (1971), Vol. II, Chapter 8, (8.14), and Tkachuk (1975, 1977) cited above. These refined results will be given in Corollaries 1.1.1 - 1.1.3 below. However, we should first obtain upper estimates for

$$\mathbf{P}\{S_n > y\} \qquad \text{and} \qquad \mathbf{P}\{\max_{1 \le k \le n} S_k \ge y\}.$$

A number of such estimates valid under various constraints on the tail behavior of function $F(\cdot)$ can be found in Petrov (1975a, 1975b) and S.V.Nagaev (1979).

We first give upper estimates for $\mathbf{P}\{S_n > y\}$.

Proposition 1.1.1. Let Condition (0.10) be fulfilled with $\beta \in (0,1) \cup (1,2)$, and $\mathbf{E}X_i = 0$

if $\beta \in (1,2)$. Then there exists a positive constant K such that for any integer $n \geq 1$ and for any real $x > 0$,

(1.1.15) $\quad \mathbf{P}\{|S_n| \geq x\} \leq K \cdot n \cdot x^{-\beta}$.

Note that in the range of deviations $x \geq n^{1/\beta+\kappa}$, where $\kappa > 0$ is any fixed real, estimate (1.1.15) follows from Theorems 1.1 - 1.2 of S.V. Nagaev (1979).

Remark 1.1.1. Note that (1.1.15) results in the following estimate which will be used in Chapter 3 in the proof of Proposition 3.1.1:

> There exists function $X(\cdot)$ from $\mathbf{R}_+^1$ to $\mathbf{R}_+^1$, such that for any integer $n \geq 1$, for any real $\varepsilon > 0$, and for $x \geq X(\varepsilon) \cdot n^{1/\beta}$, we have $\mathbf{P}\{|S_n| \geq x\} \leq \varepsilon$. The function $X(\varepsilon)$ can be chosen as $K^{1/\beta} \cdot \varepsilon^{-1/\beta}$.

Proof of Proposition 1.1.1. It is easily seen that for any positive x,

$$\mathbf{E}(1 - \exp\{-(S_n/x)^2\}) \geq \int_{\{|z| \geq x\}} (1 - \exp\{-z^2/x^2\}) \cdot dF_{S_n}(z).$$

Hence,

$$\mathbf{P}\{|S_n| \geq x\} \leq \frac{\mathbf{E}(1 - \exp\{-(S_n/x)^2\})}{1 - 1/e}.$$

Obviously,

$$\text{(1.1.16)} \qquad \mathbf{E}\exp\{-(S_n/x)^2\} = \int_{-\infty}^{\infty} e^{-t^2} \cdot dF_{S_n/x}(t).$$

Now, note that $\exp\{-t^2\}$ is the characteristic function of the normal distribution with mean zero and variance two. Therefore,

$$\mathbf{E}\exp\{-(S_n/x)^2\} = \int_{-\infty}^{\infty} \Big(\int_{-\infty}^{\infty} \frac{e^{itu}}{\sqrt{4\pi}} \cdot e^{-u^2/4} \cdot du\Big) \cdot dF_{S_n/x}(t)$$

$$\text{(1.1.17)} \quad = \int_{-\infty}^{\infty} \Big(\int_{-\infty}^{\infty} e^{itu} \cdot dF_{S_n/x}(t)\Big) \cdot \frac{e^{-u^2/4}}{\sqrt{4\pi}} \cdot du = \int_{-\infty}^{\infty} f_{X_1}(u/x)^n \cdot \frac{e^{-u^2/4}}{\sqrt{4\pi}} \cdot du,$$

where

$$f_{X_1}(v) := \mathbf{E}\exp\{ivX_1\}$$

is the characteristic function of X_1; the interchange of the order of integration is justified by Fubini's theorem. It follows from (1.1.16) and (1.1.17) that

$$(1.1.18) \qquad 1 - \mathbf{E}\exp\{-(S_n/x)^2\} = \int_{-\infty}^{\infty} (1 - f_{X_1}(u/x)^n) \cdot \frac{e^{-u^2/4}}{\sqrt{4\pi}} \cdot du.$$

It is not difficult to show that if the conditions of Proposition 1.1.1 are fulfilled, the characteristic function $f_{X_1}(v)$ admits the following representation as $|v| \to 0$:

$$f_{X_1}(v) = 1 + O(|v|^\beta)$$

(cf. Boas (1967) Theorem 1 for $\beta \in (0,1)$). Note that for $\beta \in (1,2)$, and $\mathbf{E}X_i = 0$, the hereabove representation was obtained in Binmore and Stratton (1969), Remark 1 to Theorem 1. Similar results are also given in Pitman (1968). Hence, there exist positive constants δ and K_2, such that if $|v| \le \delta$, then

$$|f_{X_1}(v) - 1| \le K_2 \cdot |v|^\beta.$$

This implies that if $|u| \le \delta \cdot x$ then

$$|1 - f_{X_1}(u/x)^n| \le K_2 \cdot n \cdot x^{-\beta} \cdot |u|^\beta.$$

Therefore,

$$(1.1.19) \qquad \begin{aligned} &\int_{|u| \le \delta x} (1 - f_{X_1}(u/x)^n) \cdot \frac{e^{-u^2/4}}{\sqrt{4\pi}} \cdot du \\ &\le K_2 \cdot n \cdot x^{-\beta} \cdot \int_{-\infty}^{\infty} |u|^\beta \cdot \frac{e^{-u^2/4}}{\sqrt{4\pi}} \cdot du = K_3 \cdot n \cdot x^{-\beta}. \end{aligned}$$

On the other hand, it is quite easy to see that the absolute value of the integral over $u : |u| > \delta \cdot x$ does not exceed

$$\pi^{-1/2} \cdot \exp\{-\delta^2 \cdot x^2 / 4\}.$$

Together with (1.1.18) and (1.1.19) this yields

$$\mathbf{E}\,(1 - \exp\{-(S_n / x)^2\}) \le K_3 \cdot n \cdot x^{-\beta} + \pi^{-1/2} \cdot \exp\{-\delta^2 \cdot x^2 / 4\}.$$

The rest of the proof is trivial. □

Proposition 1.1.2. Let Condition (0.10) be fulfilled with $\beta > 2$, $\mathbf{E}X_i = 0$, and $\mathbf{V}X_i = 1$. Then there exist positive constants K_1 and K_2, such that for any integer $n \ge 1$, and for any real $x > 0$,

$$\mathbf{P}\{|S_n| \geq x\} \leq K_1 \cdot n \cdot x^{-\beta} + \exp\{-K_2 \cdot x^2 / n\}.$$

Proof of Proposition 1.1.2 is obtained by applying Corollary 1.7 of S.V. Nagaev (1979). □

Now, let us derive upper estimates for

$$\mathbf{P}\{\max_{1 \leq k \leq n} S_k \geq y\}.$$

Lemma 1.1.3 (cf. Theorem 3 of Petrov (1975b)). Let Condition (0.10) be fulfilled with $\beta \in (0,1) \cup (1,2)$, and $\mathbf{E}X_i = 0$ if $\beta \in (1,2)$. Then there exists a positive constant K, such that for any integer $n \geq 1$, for any real $\theta > 0$ and $\kappa > 0$; and for $y \geq ((K \cdot (1+\theta)/\theta)^{1/\beta} + \kappa) \cdot n^{1/\beta}$,

$$\mathbf{P}\{\max_{1 \leq k \leq n} S_k \geq y\} \leq (1+\theta) \cdot \mathbf{P}\{S_n \geq y - (K \cdot (1+\theta)/\theta)^{1/\beta} \cdot (n-1)^{1/\beta}\}.$$

Lemma 1.1.4 (cf. Petrov (1975a) Chapter III, Theorem 12). Let Condition (0.10) be fulfilled with $\beta > 2$, $\mathbf{E}X_i = 0$, and $\mathbf{V}X_i = 1$. Then for any integer $n \geq 1$ and for any real $y > 0$,

$$\mathbf{P}\{\max_{1 \leq k \leq n} S_k \geq y\} \leq 2 \cdot \mathbf{P}\{S_n \geq y - \sqrt{2n}\}.$$

At this conclusive stage of this section, we demonstrate how to refine relationship (0.2) by the use of Theorem 1.1.1, Propositions 1.1.1 - 1.1.2, and Lemmas 1.1.3 - 1.1.4.

Corollary 1.1.1. Let Conditions (0.1) and (0.10) be fulfilled with $\beta = \alpha_1 \in (0,1) \cup (1,2)$; $\mathbf{E}X_i = 0$ if $\alpha_1 > 1$. Then there exist positive constants K_1 and K_2, such that for any integer $n \geq 1$ with $y \geq K_1 \cdot n^{1/\alpha_1}$,

$$|\mathbf{P}\{S_n > y\} - n \cdot c_{\alpha_1} \cdot y^{-\alpha_1}| \leq K_2 \cdot (n \cdot y^{-\alpha_1})^{1+1/(1+\alpha_1)} + n \sup_{x \geq y/3} |1 - F(x) - c_{\alpha_1} \cdot x^{-\alpha_1}|.$$

Corollary 1.1.1 follows from (1.1.1) if we set

$$\phi(n,y) := y \cdot (y/n^{1/\alpha_1})^{-\alpha_1/(1+\alpha_1)}$$

and then apply Proposition 1.1.1 and Lemma 1.1.3. □

Corollary 1.1.2. Let Conditions (0.1) and (0.10) be fulfilled with $\beta = \alpha_1 > 2$, $\mathbf{E}X_i = 0$, and $\mathbf{V}X_i = 1$. Then there exist positive constants K_1, K_2, and K_3, such that for any integer

$n \geq 1$ with $y \geq (K_1 \cdot n \cdot \log n)^{1/2}$,

$$|\mathbf{P}\{S_n > y\} - n \cdot c_{\alpha_1} \cdot y^{-\alpha_1}| \leq K_2 \cdot (n \cdot y^{-\alpha_1}) \cdot \{(y/\sqrt{n})^{-\beta/(1+\beta)}$$

$$+ \exp(-K_3 \cdot (y^2/n)^{1/(1+\beta)})\} + n \cdot \sup_{x \geq y/3} |1 - F(x) - c_{\alpha_1} \cdot x^{-\alpha_1}|.$$

Corollary 1.1.2 follows from (1.1.1) if we set

$$\phi(n,y) := y \cdot (y/\sqrt{n})^{-\alpha_1/(1+\alpha_1)}$$

and then apply Proposition 1.1.2 and Lemma 1.1.4. □

Corollary 1.1.3. Let Conditions (0.1) and (0.1′) be fulfilled with $\alpha_1 = 2$, and $\mathbf{E}X_i = 0$. Then there exist positive constants K_1 and K_2, such that for any integer $n \geq 2$ with $y \geq (K_1 \cdot n \cdot \log n \cdot \log\log (e \cdot n))^{1/2}$,

$$|\mathbf{P}\{S_n > y\} - n \cdot c_2 \cdot y^{-2}| \leq K_2 \cdot n \cdot y^{-2} \cdot n^{1/3} \cdot y^{-2/3} + n \cdot \sup_{x \geq y/3} |1 - F(x) - c_2 \cdot x^{-2}|.$$

Corollary 1.1.3 follows from (1.1.1) if we set $\phi(n,y) := (ny)^{1/3}$ and then apply relationship (0.2′) (which is due to Tkachuk (1975)) and Theorem 3 of Petrov (1975b). □

1.2 Asymptotic Expansions of the Probabilities of Large Deviations of S_n Taking into Account the Case when Two Summands are Comparable with the Sum.

In this section, we construct asymptotic expansions for $\mathbf{P}\{S_n > y\}$ under the assumptions that the right-hand tail of $F(x)$ admits an expansion in negative powers of x as $x \to \infty$, and the asymptotics of the left-hand tail of $F(x)$ is known up to equivalence (i.e., Conditions (0.5) and (0.1′) are fulfilled). Theorem 1.2.1 is related to the case $\alpha_1 \in (0,1) \cup (1,2)$, in which the weak convergence to the non-normal stable law with index α_1 holds, whereas Theorem 1.2.2 is related to the case of non-integer $\alpha_1 > 2$, in which the weak convergence to the normal law holds. The results of the both theorems are of the second polar type of the limiting behavior of the probabilities of large deviations of S_n, due to the fact that in the case $\alpha_1 > 2$ we do not consider the full range of large deviations, but only that where large deviations of S_n are generated mainly by one large summand X_i. However, the asymptotic expansion of Theorem 1.2.2 is more cumbersome (compare to that of Theorem 1.2.1), though the initial terms of the expansions of the both theorems are the same, namely,

$$n \cdot \sum_{i=1}^{t} c_{\alpha_i} \cdot y^{-\alpha_i},$$

in the considered ranges of deviations. Such cumbersomeness originates from the fact that the nature of the formation of large deviations in the case of the normal limit law is more complicated than that in the case of a non-normal stable limit law. This is seen not only in the main terms of the asymptotics of the probabilities of these deviations (compare (0.2) to (0.3)), but in some refining terms as well.

Let us state the main results of this section.

Theorem 1.2.1. Let Conditions (0.5) and (0.1′) be fulfilled with $\alpha_1 \in (0,1) \cup (1,2)$ and $EX_1 = 0$ if $\alpha_1 > 1$. Then

$$\mathbf{P}\{S_n > y\} = n \cdot \sum_{i=1}^{t} c_{\alpha_i} \cdot y^{-\alpha_i} - \binom{n}{2} \cdot \frac{\Gamma(1-\alpha_1)^2}{\Gamma(1-2\alpha_1)} \cdot c_{\alpha_1}^2 \cdot y^{-2\alpha_1}$$

$$- \; 2 \cdot \binom{n}{2} \cdot \frac{\Gamma(1-\alpha_1) \cdot \Gamma(2\alpha_1)}{\Gamma(\alpha_1)} \cdot c_{\alpha_1} \cdot d_{\alpha_1} \cdot y^{-2\alpha_1} + r_1(n,y) + r_2(n,y).$$

Here we use the fact that the analytic continuation of the gamma function $\Gamma(\cdot)$ onto $\mathbf{C} \setminus \{0; -1; -2; \ldots\}$ can be defined as $\Gamma(z) = \Gamma(z+1)/z$ for $Re\, z$ being negative non-integer, and $\Gamma(z) = \infty$ for $Re\, z$ being a non-positive integer; and the remainders $r_1(\cdot,\cdot)$ and $r_2(\cdot,\cdot)$ are such that

i) there exist positive constants K_1 and K_2, such that for any integer $n \geq 1$ and for $y \geq K_1 \cdot n^{1/\alpha_1}$,

$$|r_1(n,y)| \leq K_2 \cdot n \cdot \sup_{x \geq y/5} |1 - F(x) - \sum_{i=1}^{t} c_{\alpha_i} \cdot x^{-\alpha_i}|;$$

ii) there exists function $K(\cdot)$ from $\mathbf{R}_+^1$ into $\mathbf{R}_+^1$, such that for any real $\varepsilon > 0$, for any integer $n \geq 1$, and for $y \geq K(\varepsilon) \cdot n^{1/\alpha_1}$,

$$|r_2(n,y)| \leq \epsilon \cdot (n \cdot y^{-\alpha_1})^2.$$

An analogous result is also valid for non-integer values of $\alpha_1 > 2$. To this end, let us first introduce the following additional Condition (**C**), which is due to H. Cramér:

(**C**) $\qquad \limsup_{|t|\to\infty} |f_{X_i}(t)| < 1.$

Theorem 1.2.2. Let Conditions (0.5) and (0.1′) be fulfilled with non-integer $\alpha_1 > 2$, $\mathbf{E}X_i = 0$, and $\mathbf{V}X_i = 1$. In the case $\alpha_1 > 3$ we also assume the fulfilment of Condition (**C**). Then,

$$\mathbf{P}\{S_n > y\} = n\cdot\sum_{i=1}^{l} c_{\alpha_i}\cdot y^{-\alpha_i}\cdot\left(1 + \sum_{m=2}^{M(\alpha_i,\kappa)} (-1)^m\cdot\binom{-\alpha_i}{m}\cdot y^{-m}\cdot (n-1)^{m/2}\right.$$

$$\left.\cdot \int_{-\infty}^{\infty} v^m\cdot d\left\{\Phi(v) + \sum_{v=1}^{m\wedge[\alpha_i]-2} Q_v(v)/(n-1)^{v/2}\right\}\right) - \binom{n}{2}\cdot\frac{\Gamma(1-\alpha_1)^2}{\Gamma(1-2\alpha_1)}\cdot c_{\alpha_1}^2$$

$$\cdot\ y^{-2\alpha_1} - 2\cdot\binom{n}{2}\cdot\frac{\Gamma(1-\alpha_1)\cdot\Gamma(2\alpha_1)}{\Gamma(\alpha_1)}\cdot c_{\alpha_1}\cdot d_{\alpha_1}\cdot y^{-2\alpha_1} + r_1(n,y) + r_2^{(\kappa)}(n,y).$$

Hereinafter, the sum over v involving the functions $Q_v(\cdot)$ is assumed to be zero if the upper summation index is zero, the formulas for computing the functions $Q_v(\cdot)$ can be found in Petrov (1975a), Chapter VI, (1.13), $\binom{-\alpha_i}{m}$ is the coefficient under t^m in the Taylor expansion of function $(1+t)^{-\alpha_i}$ in powers of t at the neighborhood of zero:

$$\binom{-\alpha_i}{0} = 1,$$

and

$$\binom{-\alpha_i}{m} = (-1)^m\cdot\frac{\alpha_i\cdot(\alpha_i+1)\cdot\ldots\cdot(\alpha_i+m-1)}{m!}$$

for any integer $m \geq 1$, $M(\alpha_i,\kappa) = [\alpha_i + (\alpha_i - 2)/(2\kappa)]$, where $[x]$ denotes the integer part of x. Also, the remainder $r_1(\cdot,\cdot)$ is the same as in Theorem 1.2.1 and the remainder $r_2^{(\kappa)}(\cdot,\cdot)$ is such that there exists function $K(\cdot,\cdot)$ from $\mathbf{R}_+^1 \otimes \mathbf{R}_+^1$ into $\mathbf{R}_+^1$, such that for any real $\varepsilon > 0$, for any real $\kappa > 0$, for any integer $n \geq 1$, and for $y \geq K(\varepsilon,\kappa)\cdot n^{1/2+\kappa}$,

$$|r_2^{(\kappa)}(n,y)| \le \epsilon \cdot (n \cdot y^{-\alpha_1})^2.$$

Note that all the integrals from $-\infty$ to $+\infty$, contained in the asymptotic expansion, can also be expressed in terms of moments of X_i's and powers of $(n - 1)$.

The proof of Theorem 1.2.1 is analogous to that of Theorem 1.2.2, but much simpler and therefore omitted. In addition, this proof is duplicated in Chapter 3 (cf. Theorem 3.1.1 therein).

We first formulate the auxiliary results on the probabilities of normal deviations with non-uniform estimates of remainders contained in the works by Bikjalis (1966) and Osipov (1972). These results will be used in the proof of Theorem 1.2.2, and also in the proof of Theorems 1.3.1′ and 1.3.2 of the next section.

Set $F_n(x) := \mathbf{P}\{S_n \le x \cdot \sqrt{n}\}$; let $\{\delta_n\}$ denote a sequence of positive numbers (which can be different in each specific case) tending to zero as $n \to \infty$.

Proposition 1.2.1 (cf. Theorem 4 of Bikjalis (1966)). Let $\{X_n, n \ge 1\}$ be i.i.d.r.v.'s having $\mathbf{E}X_i = 0$, and $\mathbf{V}X_i = 1$. Then for any integer $n \ge 1$, and for any real x,

$$|F_n(x) - \Phi(x)| \le \frac{K}{(1+|x|)^3 \cdot \sqrt{n}} \cdot \int_0^{(1+|x|)\cdot\sqrt{n}} \Big(\int_{|u|>v} u^2 \cdot dF(u)\Big) \cdot dv.$$

Proposition 1.2.2 (cf. Corollary 2 from Theorem 3 of Bikjalis (1966)). Let $\{X_n, n \ge 1\}$ be i.i.d. non-lattice r.v.'s having $\mathbf{E}X_i = 0$, $\mathbf{V}X_i = 1$, and $\mathbf{E}|X_i|^3 < \infty$. Then for any integer $n \ge 1$, and for any real x,

$$\Big|F_n(x) - \Phi(x) - \frac{\mathbf{E}X_i^3}{6\cdot\sqrt{2\pi n}} \cdot (1 - x^2) \cdot e^{-x^2/2}\Big| \le \frac{\delta(n)}{(1+|x|)^3 \cdot \sqrt{n}}.$$

Proposition 1.2.3 (cf. Theorem 1 of Osipov (1972)). Let $\{X_n, n \ge 1\}$ be i.i.d. r.v.'s satisfying Condition (C) and having $\mathbf{E}X_i = 0$, $\mathbf{V}X_i = 1$, and $\mathbf{E}|X_i|^\ell < \infty$ for some integer $\ell \ge 3$. Then for any integer $n \ge 1$, and for any real x,

$$\Big|F_n(x) - \Phi(x) - \sum_{v=1}^{\ell-2} Q_v(x)/n^{v/2}\Big| \le K \cdot \Big\{n^{-(\ell-2)/2} \cdot (1+|x|)^{-\ell}$$

$$\cdot \int_{|v| \geq \sqrt{n}(1+|x|)} |v|^{\ell} \cdot dF(v) + n^{-(\ell-1)/2} \cdot (1+|x|)^{-\ell-1} \cdot \int_{|v| < \sqrt{n}(1+|x|)} |v|^{\ell+1} \cdot dF(v)$$

$$+ \left(\sup_{|t| \geq \delta} |f_{X_i}(t)| + 1/2n \right)^n \cdot n^{k(k+1)/2} \cdot (1+|x|)^{-k-1} \Big\},$$

where $\delta = 1 / (12 \cdot \mathbf{E}|X_i|^3)$.

For an integer $k \geq 2$ set

$$\Delta_n(z,k) := F_{S_n}(z) - \Phi(z/\sqrt{n}) - \sum_{v=1}^{k-2} Q_v(z/\sqrt{n})/n^{v/2},$$

where in the case $k = 2$, the sum over v is assumed to be zero.

Then the following estimates for $|\Delta_n(z,k)|$ can be easily derived from Propositions 1.2.1 - 1.2.3.

Corollary 1.2.1. Let Condition (0.10) be fulfilled with non-integer $\beta > 2$; $\mathbf{E}X_i = 0$, and $\mathbf{V}X_i = 1$. For $\beta > 3$ we also assume the fulfilment of Condition (**C**). Then for any integer $n \geq 1$, and for any real z,

(1.2.1) $$|\Delta_n(z, [\beta])| \leq K \cdot n \cdot (n^{1/2} + |z|)^{-\beta}.$$

Corollary 1.2.2. Let $\mathbf{E}|X_i|^\beta < \infty$ for some $\beta > 2$, $\mathbf{E}X_i = 0$, and $\mathbf{V}X_i = 1$. In the cases $\beta = 3$ and $\beta > 3$ we also assume that either r.v.'s $\{X_n, n \geq 1\}$ are not lattice variables or Condition (**C**) is fulfilled, respectively. Then for any integer $n \geq 1$, and for any real z,

(1.2.2) $$|\Delta_n(z, [\beta])| \leq \delta(n) \cdot n \cdot (\sqrt{n} + |z|)^{-\beta}.$$

It is easily seen that estimate (1.2.1) of Corollary 1.2.1 is asymptotically exact in the sense of dependence on n and z. In fact, letting both n and $z/\sqrt{n}$ tend to infinity in (1.2.1) we get that

$$\mathbf{P}\{S_n > z\} = 1 - \Phi(z/\sqrt{n}) - \sum_{v=1}^{[\beta]-2} Q_v(z/\sqrt{n})/n^{v/2} + O(n \cdot z^{-\beta}).$$

It only remains to note that the remainder, $O(n \cdot z^{-\beta})$, is $O(n \cdot \mathbf{P}\{X_1 > y\})$ and then compare it to (0.3).

Now, let us note that the fulfilment of (0.5) and (0.1′) with a certain non-integer $\alpha_1 > 2$ implies the fulfilment of (0.10). Hence, in order to prove Theorem 1.2.2 we may apply Corollary 1.2.1 with $\beta = \alpha_1$.

Proof of Theorem 1.2.2 is essentially carried out by the direct probabilistic method described in the Introduction and is analogous to those of Theorem 1.1.1 and Corollary 1.1.2 of the previous section. For $n = 1$ the assertion of the theorem follows from Condition (0.5). Hence, we can assume that $n \geq 2$.

First, let us apply relationship (1.1.2) with $u = y/5$. Then Lemma 1.1.1 yields

$$\mathbf{P}\{S_n > y,\ \max(X_1,\dots,X_n) \leq y/5\} \leq \mathbf{P}\{\max_{1\leq k\leq n} S_k > y/5\}^3. \tag{1.2.3}$$

We drop the terms of the alternating sum on the right-hand side of (1.1.2) beginning with the third one. Then Lemma 1.1.2 implies that

$$\begin{aligned} &\left|\sum_{m=3}^{n} (-1)^m \cdot \binom{n}{m} \cdot \mathbf{P}\{S_n > y,\ X_{n-m+1} > y/5,\dots,X_n > y/5\}\right| \\ &\leq \binom{n}{3} \cdot \mathbf{P}\{X_1 > y/5\}^3. \end{aligned} \tag{1.2.4}$$

Combining (1.1.2) with $u = y/5$, (1.2.3), and (1.2.4) we obtain the result that for any integer $n \geq 2$ and for any real $y > 0$,

$$\begin{aligned} &\left|\mathbf{P}\{S_n > y\} - n\cdot\mathbf{P}\{S_n > y, X_n > y/5\} + \binom{n}{2}\cdot\mathbf{P}\{S_n > y, X_{n-1} > y/5, X_n > y/5\}\right| \\ &\leq \mathbf{P}\{\max_{1\leq k\leq n} S_k > y/5\}^3 + \binom{n}{3}\cdot\mathbf{P}\{X_1 > y/5\}^3. \end{aligned}$$

Let us estimate the leftmost and the rightmost terms on the right-hand side of this inequality by the use of Lemma 1.1.4 and Corollary 1.1.2; and expansion given by (0.5), respectively. We obtain that for any integer $n \geq 2$ and for $y \geq K_1 \cdot (n\cdot\log n)^{1/2}$,

$$\begin{aligned} &\left|\mathbf{P}\{S_n>y\} - n\cdot\mathbf{P}\{S_n> y, X_n>y/5\} + \binom{n}{2}\cdot\mathbf{P}\{S_n>y, X_{n-1}>y/5, X_n>y/5\}\right| \\ &\leq K_2 \cdot (n\cdot y^{-\alpha_1})^3. \end{aligned} \tag{1.2.5}$$

The probability contained in the rightmost term on the left-hand side of (1.2.5) admits the following estimate: there exists a function $K(\cdot)$ from $\mathbf{R}_+^1$ into $\mathbf{R}_+^1$ such that for any integer $n \geq 2$, for any real $\varepsilon > 0$, and for $y \geq K(\varepsilon) \cdot (n \cdot \log n)^{1/2}$,

(1.2.6)
$$|\mathbf{P}\{S_n>y, X_{n-1}>y/5, X_n>y/5\} - c_{\alpha_1}^2\cdot\Big((4/25)^{-\alpha_1} + \int_{1/5}^{4/5}(1-t)^{-\alpha_1}\cdot d(-t^{-\alpha_1})\Big)\cdot y^{-2\cdot\alpha_1}| \le \epsilon\cdot y^{-2\cdot\alpha_1}.$$

The idea of the derivation of this estimate is as follows. The independence of S_{n-2}, X_{n-1}, and X_n implies that

$$\mathbf{P}\{S_n>y, X_{n-1}>y/5, X_n>y/5\} = \mathbf{P}\{S_{n-2}>3y/5\}\cdot\mathbf{P}\{X_{n-1}>y/5\}\cdot\mathbf{P}\{X_n>y/5\}$$
$$+ \int_{-\infty}^{3y/5} \mathbf{P}\{X_{n-1}>y/5, X_n>y/5, X_{n-1}+X_n>y-z\}\cdot d\mathbf{F}_{S_{n-2}}(z).$$

After simple transformations we obtain that

(1.2.7)
$$\mathbf{P}\{S_n>y, X_{n-1}>y/5, X_n>y/5\} = \mathbf{P}\{X_{n-1}>y/5\}\cdot\mathbf{P}\{X_n>y/5\}$$
$$- \int_{-\infty}^{3y/5} \mathbf{P}\{X_{n-1}>y/5, X_n>y/5, X_{n-1}+X_n\le y-z\}\cdot d\mathbf{F}_{S_{n-2}}(z).$$

The first term on the right-hand side of (1.2.7) is properly approximated by

$$\Big(\sum_{i=1}^{l} c_{\alpha_i}\cdot (y/5)^{-\alpha_i}\Big)^2.$$

In turn, due to (0.5), the integrand in the second term on the right-hand side of (1.2.7) is properly approximated by

$$\int_{y/5}^{4y/5-z}\Big(\sum_{i=1}^{l} c_{\alpha_i}\cdot(y/5)^{-\alpha_i} - (y-z-v)^{-\alpha_i}\Big)\cdot d\Big(-\sum_{i=1}^{l} c_{\alpha_i}\cdot v^{-\alpha_i}\Big)$$

in the full range of integration over $z \in (-\infty, 3y/5]$. In addition, the function $\mathbf{F}_{S_{n-2}}(z)$ is properly approximated in terms of the expressions

$$(n-2)\cdot d_{\alpha_1}\cdot|z|^{-\alpha_1}$$

for $z \le -(K_1\cdot n\cdot \log n)^{1/2}$ and

$$1-(n-2)\cdot c_{\alpha_1}\cdot z^{-\alpha_1}$$

for $z \geq (K_2 \cdot n \cdot \log n)^{1/2}$, by (0.5), (0.1′), Corollary 1.1.2, and its left-hand analog. The derivation of estimate (1.2.6) relies on the utilization of these approximations, and is similar to the proofs of Lemmas 3.1.8 and 3.1.9 of Section 3.1 (but much simpler), and therefore is omitted. For a special case $n \to \infty$, $y \geq K \cdot n^{1/2+\kappa}$, where $\kappa > 0$ is any fixed real, the required representation for the rightmost term on the left-hand side of (1.2.5) is relatively easy to obtain by applying Theorem 2′ and Remark 4 of Godovan'chuk (1981):

$$\binom{n}{2}\cdot \mathbf{P}\{S_n > y,\ X_{n-1} > y/5,\ X_n > y/5\}$$
$$\sim \binom{n}{2}\cdot c_{\alpha_1}^2 \cdot \left((4/25)^{-\alpha_1} + \int_{1/5}^{4/5} (1-t)^{-\alpha_1}\cdot d(-t^{-\alpha_1})\right)\cdot y^{-2\alpha_1}.$$

We now estimate the middle term on the left-hand side of (1.2.5) more precisely by the use of expansion (0.5) and Corollary 1.1.2. By analogy with relationships (1.1.8) - (1.1.11) of the previous section we conclude that for any positive y,

$$\Big|\mathbf{P}\{S_n > y, X_n > y/5\} - \sum_{i=1}^{\ell} c_{\alpha_i}\cdot\Big((y/5)^{-\alpha_i}\cdot \mathbf{P}\{S_{n-1} > 4y/5\}$$

(1.2.8)

$$+ \int_{-\infty}^{4y/5} (y-z)^{-\alpha_i}\cdot dF_{S_{n-1}}(z)\Big)\Big| \leq K\cdot \sup_{x\geq y/5}\Big|1-F(x)-\sum_{i=1}^{\ell} c_{\alpha_i}\cdot x^{-\alpha_i}\Big|.$$

Note that the expression on the right-hand side of (1.2.8) multiplied by n will be included in the remainder $r_1(n,y)$. Now we proceed with the estimation of the first (and the main) summand of the sum over i on the left-hand side of (1.2.8) (note that the other $\ell - 1$ summands are estimated following along the same lines, but in a simpler manner):

(1.2.9) $$c_{\alpha_1}\cdot\Big((y/5)^{-\alpha_1}\cdot \mathbf{P}\{S_{n-1} > 4y/5\} + \int_{-\infty}^{4y/5} (y-z)^{-\alpha_1}\cdot dF_{S_{n-1}}(z)\Big).$$

The first term of the sum within the braces is properly approximated by

(1.2.10) $$(n-1)\cdot c_{\alpha_1}\cdot (4/25)^{-\alpha_1}\cdot y^{-2\alpha_1},$$

due to Corollary 1.1.2. In order to estimate the integral contained in (1.2.9), we first take the Taylor expansion of the integrand $(y - z)^{-\alpha_1}$ over powers of z at the neighborhood of zero. Then

$$\begin{aligned}(1.2.11)\qquad &\int_{-\infty}^{4y/5} (y-z)^{-\alpha_1}\cdot dF_{S_{n-1}}(z) = \sum_{m=0}^{[\alpha_1]} (-1)^m\cdot\binom{-\alpha_1}{m}\cdot y^{-\alpha_1-m}\cdot \int_{-\infty}^{4y/5} z^m\cdot dF_{S_{n-1}}(z)\\ &+ \int_{-\infty}^{4y/5} \Big((y-z)^{-\alpha_1} - \sum_{m=0}^{[\alpha_1]} (-1)^m\cdot\binom{-\alpha_1}{m}\cdot y^{-\alpha_1-m}\cdot z^m\Big)\cdot dF_{S_{n-1}}(z).\end{aligned}$$

Let us continue the transformation of the integral from (1.2.9). It is easily seen that for any non-negative integer $m \le [\alpha_1]$,

$$\int_{-\infty}^{4y/5} z^m\cdot dF_{S_{n-1}}(z) = \mathbf{E}S_{n-1}^m - \int_{4y/5}^{+\infty} z^m\cdot dF_{S_{n-1}}(z).$$

Expressing the moments of the sum S_{n-1} in terms of the functions $Q_v(\cdot)$ by the use of Theorem 1 of von Bahr (1965) we obtain that the leftmost sum on the right-hand side of (1.2.11) is equal to

$$\begin{aligned}(1.2.12)\qquad &1+\sum_{m=2}^{[\alpha_1]} (-1)^m\cdot\binom{-\alpha_1}{m}\cdot(n-1)^{m/2}\cdot y^{-\alpha_1-m}\cdot \int_{-\infty}^{\infty} v^m\cdot d\{\Phi(v) + \sum_{v=1}^{m-2}\frac{Q_v(v)}{(n-1)^{v/2}}\}\\ &- \sum_{m=0}^{[\alpha_1]} (-1)^m\cdot\binom{-\alpha_1}{m}\cdot y^{-\alpha_1-m}\cdot \int_{4y/5}^{+\infty} z^m\cdot d(F_{S_{n-1}}(z) - 1).\end{aligned}$$

We estimate the integrals contained in the rightmost sum in (1.2.12) by means of the integration by parts with the subsequent approximation of the difference

$$F_{S_{n-1}}(z) - 1 \qquad \text{by} \qquad -(n-1)\cdot c_{\alpha_1}\cdot z^{-\alpha_1}$$

(this approximation is justified by Corollary 1.1.2). In particular, that corollary implies that there exists a function $K(\cdot)$ from $\mathbf{R}_+^1$ to $\mathbf{R}_+^1$, such that for any real $\varepsilon > 0$, for any integer $n \ge 2$, for any non-negative integer $m \le [\alpha_1]$, and for $y \ge K(\varepsilon)\cdot(n\cdot\log n)^{1/2}$,

$$|\int_{4y/5}^{+\infty} z^m \cdot d(F_{S_{n-1}}(z) - 1) - \frac{\alpha_1}{\alpha_1 - m} \cdot (\frac{4}{5})^{m-\alpha_1} \cdot (n-1) \cdot y^{m-\alpha_1}| \tag{1.2.13}$$
$$\leq \epsilon \cdot n \cdot y^{-\alpha_1}.$$

Up till now we have used only the direct probabilistic method, but in order to investigate the rightmost integral on the right-hand side of (1.2.11), we also should apply non-uniform estimates of remainders in the theorems on normal deviations given in Corollary 1.2.1.

Let us choose a function $\bar{K}(\cdot)$ such that for $y \geq \bar{K}(\varepsilon) \cdot ((n-1) \cdot \log(n-1))^{1/2}$,

$$|\mathbf{P}\{S_{n-1} < -y\} - (n-1) \cdot d_{\alpha_1} \cdot y^{-\alpha_1}| \vee |\mathbf{P}\{S_{n-1} > y\} - (n-1) \cdot c_{\alpha_1} \cdot y^{-\alpha_1}| \tag{1.2.14}$$
$$\leq \epsilon \cdot n \cdot y^{-\alpha_1}$$

(the existence of a function with this property follows from Corollary 1.1.2 and its left-hand analog). Set $R(\varepsilon,n) = \bar{K}(\varepsilon) \cdot ((n-1) \cdot \log(n-1))^{1/2}$ and break up our integral,

$$\int_{-\infty}^{4y/5} ((y-z)^{-\alpha_1} - \sum_{m=0}^{[\alpha]} (-1)^m \cdot \binom{-\alpha_1}{m} y^{-\alpha_1 - m} \cdot z^m) \cdot d\mathbf{F}_{S_{n-1}}(z), \tag{1.2.15}$$

into the integrals from $-\infty$ to $-R(\varepsilon,n)$ (I_1); from $-R(\varepsilon,n)$ to $R(\varepsilon,n)$ (I_2), and from $R(\varepsilon,n)$ to $4y/5$ (I_3). We estimate these integrals separately.

We proceed with the study of I_2. To this end, we construct the Taylor expansion in powers of z with Lagrange's form for the remainder for the function $(y-z)^{-\alpha_1}$ at the neighborhood of zero. Then we get that for any non-negative integer M, and for $|z| \leq R(\varepsilon,n)$,

$$|(y-z)^{-\alpha_1} - \sum_{m=0}^{M} (-1)^m \cdot \binom{-\alpha_1}{m} \cdot y^{-\alpha_1 - m} \cdot z^m| \leq K \cdot y^{-\alpha_1} \cdot (z/y)^{-M-1}, \tag{1.2.16}$$

where the constant K can depend on M and α_1. Assume that $M = M(\alpha_1,\kappa)$ and choose function $V(\cdot)$: $\mathbf{R}_+^1 \to \mathbf{R}_+^1$ satisfying the condition that for any integer $n \geq 1$ and for any real $\kappa > 0$,

$$(1.2.17) \qquad (\log n)^{\frac{M(\alpha_1,\kappa)+1}{2}} \cdot n^{-\kappa\left(1-\left\{\alpha_1+\frac{\alpha_1-2}{2\kappa}\right\}\right)} \cdot V(\kappa)^{-(M(\alpha_1,\kappa)+1-\alpha_1)} \le 1.$$

Then (1.2.16) and (1.2.17) imply that for any integer $n \ge 1$, for any real $\varepsilon > 0$ and $\kappa > 0$, for

$$y \ge \bar{K}(\epsilon)^{\frac{M(\alpha_1,\kappa)+1}{M(\alpha_1,\kappa)+1-\alpha_1}} \cdot \epsilon^{\frac{-1}{M(\alpha_1,\kappa)+1-\alpha_1}} \cdot V(\kappa) \cdot n^{1/2+\kappa},$$

and in the range of variation of z such that $|z| \le R(\varepsilon,n)$,

$$\left|(y-z)^{-\alpha_1} - \sum_{m=0}^{M(\alpha_1,\kappa)} (-1)^m \cdot \binom{-\alpha_1}{m} \cdot y^{-\alpha_1-m} \cdot z^m\right| \le \epsilon \cdot n \cdot y^{-2\alpha_1}.$$

The latter inequality yields that for any integer $n \ge 1$ and in the same range of variation of y,

$$(1.2.18) \qquad \left|\int_{-R(\epsilon,n)}^{R(\epsilon,n)} \left((y-z)^{-\alpha_1} - \sum_{m=0}^{M(\alpha_1,\kappa)} (-1)^m \cdot \binom{-\alpha_1}{m} \cdot y^{-\alpha_1-m} \cdot z^m\right) \cdot dF_{S_{n-1}}(z)\right| \le \epsilon \cdot n \cdot y^{-2\alpha_1}.$$

It is easily seen that (1.2.18) is in fact the error estimate for the absolute value of the difference between I_2 and the following expression:

$$S(\epsilon,\kappa) := \sum_{m=[\alpha_1]+1}^{M(\alpha_1,\kappa)} (-1)^m \cdot \binom{-\alpha_1}{m} \cdot y^{-\alpha_1-m} \cdot \int_{-R(\epsilon,n)}^{R(\epsilon,n)} z^m \cdot dF_{S_{n-1}}(z).$$

Hence, (1.2.18) implies that for any integer $n \ge 2$, for any real $\varepsilon > 0$ and $\kappa > 0$, and for

$$y \ge \bar{K}(\epsilon)^{\frac{M(\alpha_1,\kappa)+1}{M(\alpha_1,\kappa)+1-\alpha_1}} \cdot \epsilon^{\frac{-1}{M(\alpha_1,\kappa)+1-\alpha_1}} \cdot V(\kappa) \cdot n^{1/2+\kappa},$$

$$(1.2.19) \qquad |I_2 - S(\epsilon,\kappa)| \le \epsilon \cdot n \cdot y^{-2\alpha_1}.$$

Note that the expression on the right-hand side of (1.2.19) multiplied by n will be included in the remainder $r_2^{(\kappa)}(n,y)$.

Now, in order to investigate the behavior of the sum $S(\varepsilon,\kappa)$, we replace the function $F_{S_{n-1}}(z)$

contained in the integral on the left-hand side of (1.2.18) by the corresponding Edgeworth expansion estimating the error emerging from this change by the use of Corollary 1.2.1. Keeping in mind the notation introduced above Corollary 1.2.1 we obtain that

$$S(\epsilon,\kappa) := \sum_{m=[\alpha_1]+1}^{M(\alpha_1,\kappa)} (-1)^m \cdot \binom{-\alpha_1}{m} \cdot y^{-\alpha_1-m} \cdot \Bigg(\int_{-R(\epsilon,n)}^{R(\epsilon,n)} z^m \cdot d\Big(\Phi\Big(\frac{z}{\sqrt{n-1}}\Big) \tag{1.2.20}$$
$$+ \sum_{\nu=1}^{[\alpha_1]-2} \frac{Q_\nu\big(z/\sqrt{n-1}\big)}{(n-1)^{\nu/2}}\Big) + \int_{-R(\epsilon,n)}^{R(\epsilon,n)} z^m \cdot d\Delta_{n-1}(z,[\alpha_1])\Bigg).$$

In order to estimate the rightmost integral on the right-hand side of (1.2.20) we integrate it by parts and then apply Corollary 1.2.1. As a result, we obtain that for any integer $[\alpha_1] + 1 \le m \le M(\alpha_1,\kappa)$,

$$\left|\int_{-R(\epsilon,n)}^{R(\epsilon,n)} z^m \cdot \Delta_{n-1}(z,[\alpha_1])\right| \le K \cdot n \cdot \Bigg(R(\epsilon,n)^{m-\alpha_1} + m \cdot \int_{-R(\epsilon,n)}^{R(\epsilon,n)} \frac{|z|^{m-1}}{(\sqrt{n}+|z|)^{\alpha_1}} \cdot dz\Bigg)$$
$$\le K(m) \cdot n \cdot \Big(R(\epsilon,n)^{m-\alpha_1} + n^{\frac{m-\alpha_1}{2}}\Big).$$

Recall that $R(\varepsilon,n) = \bar{K}(\varepsilon) \cdot ((n-1) \cdot \log (n-1))^{1/2}$. Now, choosing

$$y \ge \bar{K}(\epsilon) \cdot \epsilon^{\frac{1}{(\alpha_1)}} \cdot \sqrt{n \cdot \log n}$$

we get that for any integer $[\alpha_1] + 1 \le m \le M(\alpha_1,\kappa)$,

$$R(\epsilon,n)^{m-\alpha_1} \cdot y^{-(m-\alpha_1)} \le \epsilon.$$

Hence, there exists a function $L(\cdot,\cdot)$ from $\mathbf{R}_+^1 \otimes \mathbf{R}_+^1$ to $\mathbf{R}_+^1$ such that for any integer $n \ge 2$, for any real $\varepsilon > 0$ and $\kappa > 0$; and for $y \ge L(\varepsilon,\kappa) \cdot (n \cdot \log n)^{1/2}$,

$$\left|\sum_{m=[\alpha_1]+1}^{M(\alpha_1,\kappa)} (-1)^m \cdot \binom{-\alpha_1}{m} \cdot y^{-\alpha_1-m} \cdot \int_{-R(\epsilon,n)}^{R(\epsilon,n)} z^m \cdot d\Delta_{n-1}(z,[\alpha_1])\right| \tag{1.2.21}$$
$$\le \epsilon \cdot n \cdot y^{-2\alpha_1}.$$

Note that the expression on the right-hand side of (1.2.21) multiplied by n will be included in the remainder $r_2^{(\kappa)}(n,y)$.

Now, let us demonstrate that the lower and the upper limits of integration in the leftmost integral contained in the mth term on the right-hand side of (1.2.20) can be replaced by $-\infty$ and $+\infty$, respectively, i.e., the errors which emerge due to these changes are absorbed into the remainder $r_2^{(\kappa)}(n,y)$. To this end, it suffices to estimate the left- and the right-hand tails, i.e. the following integrals:

$$\text{(1.2.22)} \qquad \int_{-\infty}^{-R(\epsilon,n)} z^m \cdot d\left(\Phi\left(\frac{z}{\sqrt{n-1}}\right) + \sum_{\nu=1}^{[\alpha_1]-2} \frac{Q_\nu(z/\sqrt{n-1})}{(n-1)^{\nu/2}}\right)$$

and

$$\text{(1.2.22}'\text{)} \qquad \int_{R(\epsilon,n)}^{+\infty} z^m \cdot d\left(\Phi\left(\frac{z}{\sqrt{n-1}}\right) + \sum_{\nu=1}^{[\alpha_1]-2} \frac{Q_\nu(z/\sqrt{n-1})}{(n-1)^{\nu/2}}\right).$$

Making the same change of variables $v = z/\sqrt{(n-1)}$ in both integrals we get that there exists a positive constant K, such that the absolute values of the integrals (1.2.22) and (1.2.22′) do not exceed

$$\text{(1.2.23)} \qquad (n-1)^{m/2} \cdot \exp\left\{-K \cdot \sqrt{\overline{K}(\epsilon) \cdot \log(n-1)}\right\},$$

which justifies the above made changes of the lower and the upper limits of integration for $-\infty$ and $+\infty$.

Combining estimates (1.2.19) - (1.2.21) and (1.2.23) we get that there exists a function $V(\cdot,\cdot)$ from $\mathbf{R}_+^1 \otimes \mathbf{R}_+^1$ to $\mathbf{R}_+^1$, such that for any integer $n \geq 2$, for any real $\varepsilon > 0$ and $\kappa > 0$; and for $y \geq V(\varepsilon,\kappa) \cdot n^{1/2+\kappa}$,

$$\text{(1.2.24)} \qquad \left| I_2 - \sum_{m=[\alpha_1]+1}^{M(\alpha_1,\kappa)} (-1)^m \cdot \binom{-\alpha_1}{m} \cdot y^{-\alpha_1-m} \cdot (n-1)^{\frac{m}{2}} \cdot \int_{-\infty}^{\infty} v^m \cdot d\left(\Phi(v) + \sum_{\nu=1}^{[\alpha_1]-2} \frac{Q_\nu(v)}{(n-1)^{\nu/2}}\right)\right| \leq \epsilon \cdot n \cdot y^{-2\alpha_1}.$$

Integrals I_1 and I_3 (introduced below formula (1.2.15)) are significantly simpler to estimate. The approach is as follows: we approximate function

$$F_{S_{n-1}}(z) \qquad \text{by} \qquad (n-1) \cdot d_{\alpha_1} \cdot |z|^{-\alpha_1}$$

(when considering I_1), and by

$$1-(n-1)\cdot c_{\alpha_1}\cdot z^{-\alpha_1}$$

(when considering I_3); the errors that emerge due to these approximations are easily estimated by applying Corollary 1.1.2 and its left-hand analogue. Thereupon, in the emerged integrals which approximate I_1 and I_3, respectively, changes of the upper limit of integration from $-R(\varepsilon,n)$ to zero (in the integral over negative range of variation of z) and of the lower limit of integration from $R(\varepsilon,n)$ to zero (in the integral over positive range of variation of z) should be made. An application of the Taylor expansion with Lagrange's form for the remainder to the function $(y-z)^{-\alpha_1}$ enables us to demonstrate that the two 'correcting' integrals from $-R(\varepsilon,n)$ to zero and from zero to $R(\varepsilon,n)$, multiplied by n, will be included in the remainder $r_2^{(\kappa)}(n,y)$. Thus, I_1 and I_3 are properly approximated by the following integrals:

$$\int_{-\infty}^{0}\left((y-z)^{-\alpha_1}-\sum_{m=0}^{[\alpha_1]}(-1)^m\cdot\binom{-\alpha_1}{m}\cdot y^{-\alpha_1-m}\cdot z^m\right)\cdot d((n-1)\cdot d_{\alpha_1}\cdot|z|^{-\alpha_1})$$

and

$$\int_{0}^{4y/5}\left((y-z)^{-\alpha_1}-\sum_{m=0}^{[\alpha_1]}(-1)^m\cdot\binom{-\alpha_1}{m}\cdot y^{-\alpha_1-m}\cdot z^m\right)\cdot d(-(n-1)\cdot c_{\alpha_1}\cdot z^{-\alpha_1}).$$

The subsequent change of variables $t = z/y$ turn these integrals into

$$(n-1)\cdot d_{\alpha_1}\cdot y^{-2\alpha_1}\cdot\int_{-\infty}^{0}((1-t)^{-\alpha_1}-\sum_{m=0}^{[\alpha_1]}(-1)^m\cdot\binom{-\alpha_1}{m}\cdot t^m)\cdot d|t|^{-\alpha_1}$$

and

$$(n-1)\cdot c_{\alpha_1}\cdot y^{-2\alpha_1}\cdot\int_{0}^{4y/5}((1-t)^{-\alpha_1}-\sum_{m=0}^{[\alpha_1]}(-1)^m\cdot\binom{-\alpha_1}{m}\cdot t^m)\cdot d(-t^{-\alpha_1}).$$

Further, there exists a function $W(\cdot,\cdot)$ from $\mathbb{R}_+^1$ to $\mathbb{R}_+^1$ such that for any integer $n \geq 2$, for any real $\varepsilon > 0$, and for $y \geq W(\varepsilon)\cdot(n\cdot\log n)^{1/2}$,

$$|\mathbf{I}_1 - (n-1)\cdot \mathbf{d}_{\alpha_1}\cdot y^{-2\alpha_1}\cdot \int_{-\infty}^{0}((1-t)^{-\alpha_1} - \sum_{m=0}^{[\alpha_1]}(-1)^m\cdot\binom{-\alpha_1}{m}\cdot t^m)\cdot d|t|^{-\alpha_1}|$$

(1.2.25)

$$\leq \epsilon\cdot n\cdot y^{-2\alpha_1},$$

and

$$|\mathbf{I}_3 - (n-1)\cdot \mathbf{c}_{\alpha_1}\cdot y^{-2\alpha_1}\cdot \int_{0}^{4y/5}((1-t)^{-\alpha_1} - \sum_{m=0}^{[\alpha_1]}(-1)^m\cdot\binom{-\alpha_1}{m}\cdot t^m)\cdot d(-t^{-\alpha_1})|$$

(1.2.25′)

$$\leq \epsilon\cdot n\cdot y^{-2\alpha_1}.$$

The proof of validity of inequalities (1.2.25) - (1.2.25′) is omitted. The interested reader can recover the derivation of these inequalities by considering similar (but more precise) estimates related to the case of $\alpha_1 \in (0,1) \cup (1,2)$ that are given in the proof of Theorem 3.1.1 of Chapter 3.

Combining estimates (1.2.8) - (1.2.13) with inequalities (1.2.25) for I_1, (1.2.24) for I_2, and (1.2.25′) for I_3, we get that there exists a function $K(\cdot,\cdot)$ from $\mathbf{R}_+^1 \otimes \mathbf{R}_+^1$ to $\mathbf{R}_+^1$, such that for any integer $n \geq 2$, for any real $\varepsilon > 0$ and $\kappa > 0$; and for $y \geq K(\varepsilon,\kappa)\cdot n^{1/2+\kappa}$,

$$|\mathbf{P}\{S_n > y, X_n > y/5\} - \sum_{i=1}^{l} \mathrm{c}_{\alpha_i}\cdot y^{-\alpha_i}\cdot\Big(1 + \sum_{m=2}^{M(\alpha_i,\kappa)}(-1)^m\cdot\binom{-\alpha_i}{m}\cdot y^{-m}\cdot (n-1)^{m/2}$$

$$\cdot \int_{-\infty}^{\infty} \mathrm{v}^m\cdot d\Big\{\Phi(\mathrm{v}) + \sum_{\mathrm{v}=1}^{m\wedge[\alpha_i]-2}\frac{Q_{\mathrm{v}}(\mathrm{v})}{(n-1)^{\mathrm{v}/2}}\Big\}\Big) - (n-1)\cdot\mathrm{d}_{\alpha_1}\cdot y^{-2\alpha_1}$$

$$\cdot \int_{-\infty}^{0}\Big((1-t)^{-\alpha_1} - \sum_{m=0}^{[\alpha_1]}(-1)^m\cdot\binom{-\alpha_1}{m}\cdot t^m\Big)\cdot d|t|^{-\alpha_1} - (n-1)\cdot c_{\alpha_1}\cdot y^{-2\alpha_1}$$

(1.2.26)

$$\cdot \Big((4/25)^{-\alpha_1} - \sum_{m=0}^{[\alpha_1]}(-1)^m\cdot\binom{-\alpha_1}{m}\cdot \alpha_1/(\alpha_1-m)\cdot(4/5)^{m-\alpha_1}$$

$$+ \int_{0}^{4/5}\Big((1-t)^{-\alpha_1} - \sum_{m=0}^{[\alpha_1]}(-1)^m\cdot\binom{-\alpha_1}{m}\cdot t^m\Big)\cdot d(-t^{-\alpha_1})\Big)$$

$$\leq \epsilon\cdot n\cdot y^{-2\alpha_1} + K_1\cdot \sup_{x\geq y/5}\Big|1 - F(x) - \sum_{i=1}^{l}\mathrm{c}_{\alpha_i}\cdot x^{-\alpha_i}\Big|.$$

Obviously, the assertion of Theorem 1.2.2 would easily follow from (1.2.5), (1.2.6), and (1.2.26) if we demonstrate that for any positive non-integer α, the following relationships are valid:

(1.2.27)
$$\int_{-\infty}^{0}\left((1-t)^{-\alpha}-\sum_{m=0}^{[\alpha]}(-1)^m\cdot\binom{-\alpha}{m}\cdot t^m\right)\cdot d|t|^{-\alpha}=-\frac{\Gamma(1-\alpha)\cdot\Gamma(2\alpha)}{\Gamma(\alpha)},$$

and

(1.2.27′)
$$2\cdot\left(\delta^{-\alpha}\cdot(1-\delta)^{-\alpha}-\sum_{m=0}^{[\alpha]}(-1)^m\cdot\binom{-\alpha}{m}\cdot\frac{\alpha}{\alpha-m}\cdot(1-\delta)^{m-\alpha}\right.$$
$$\left.+\int_{0}^{1-\delta}\left((1-t)^{-\alpha}-\sum_{m=0}^{[\alpha]}(-1)^m\cdot\binom{-\alpha}{m}\cdot t^m\right)\cdot d(-t^{-\alpha})\right)-\delta^{-\alpha}\cdot(1-\delta)^{-\alpha}$$
$$-\int_{0}^{1-\delta}(1-t)^{-\alpha}\cdot d(-t^{-\alpha})=-\frac{\Gamma(1-\alpha)^2}{\Gamma(1-2\alpha)},$$

where $\delta < 1/2$ is any positive (note that in our case $\delta = 1/5$).

First, let us establish (1.2.27′). Note that the expression on the left-hand side of (1.2.27′) does not depend on δ since its derivative in δ is equal to zero. After some relatively simple transformations we obtain that this expression is equal to

$$2\cdot\int_{0}^{\delta}\left((1-t)^{-\alpha}-\sum_{m=0}^{[\alpha]}(-1)^m\cdot\binom{-\alpha}{m}\cdot t^m\right)\cdot d(-t^{-\alpha})$$
$$+\int_{\delta}^{1-\delta}\left((1-t)^{-\alpha}-\sum_{m=0}^{[\alpha]}(-1)^m\cdot\binom{-\alpha}{m}\cdot t^m\right)\cdot d\left(-t^{-\alpha}+\sum_{m=0}^{[\alpha]}(-1)^m\cdot\binom{-\alpha}{m}\cdot(1-t)^m\right)$$
$$+\sum_{m=0}^{[\alpha]}(-1)^m\cdot\binom{-\alpha}{m}\cdot\frac{m}{m-\alpha}\cdot\left((1-\delta)^{m-\alpha}-\delta^{m-\alpha}\right)-\sum_{m=0}^{[\alpha]}\sum_{k=1}^{[\alpha]}k\cdot(-1)^{m+k}$$
$$\cdot\binom{-\alpha}{m}\cdot\binom{-\alpha}{k}\cdot\int_{\delta}^{1-\delta}t^m\cdot(1-t)^{k-1}\cdot dt-\sum_{m=0}^{[\alpha]}(-1)^m\cdot\binom{-\alpha}{m}\cdot\frac{\alpha}{m-\alpha}$$

$$\cdot\left((1-\delta)^{m-\alpha}-\delta^{m-\alpha}\right)-2\cdot\sum_{m=0}^{[\alpha]}(-1)^m\cdot\binom{-\alpha}{m}\cdot\frac{\alpha}{\alpha-m}\cdot(1-\delta)^{m-\alpha}+\delta^{-\alpha}\cdot(1-\delta)^{-\alpha};$$

the terms of the double sum over m and k for which $k = 0$ are assumed to be zero. It can be shown that the sum of the third, the fifth, and the sixth terms is equal to

$$-\sum_{m=0}^{[\alpha]}(-1)^m\cdot\binom{-\alpha}{m}\cdot\delta^{m-\alpha}-\sum_{m=0}^{[\alpha]}(-1)^m\cdot\binom{-\alpha}{m}\cdot\frac{\alpha+m}{\alpha-m}\cdot(1-\delta)^{m-\alpha}.$$

Hence, the left-hand side of (1.2.27′) is equal to

$$2\cdot\int_0^\delta\left((1-t)^{-\alpha}-\sum_{m=0}^{[\alpha]}(-1)^m\cdot\binom{-\alpha}{m}\cdot t^m\right)\cdot d(-t^{-\alpha})$$

$$+\int_\delta^{1-\delta}\left((1-t)^{-\alpha}-\sum_{m=0}^{[\alpha]}(-1)^m\cdot\binom{-\alpha}{m}\cdot t^m\right)\cdot d\left(-t^{-\alpha}+\sum_{m=0}^{[\alpha]}(-1)^m\cdot\binom{-\alpha}{m}\cdot(1-t)^m\right)$$

$$-\sum_{m=0}^{[\alpha]}\sum_{k=1}^{[\alpha]}k\cdot(-1)^{m+k}\cdot\binom{-\alpha}{m}\cdot\binom{-\alpha}{k}\cdot\int_\delta^{1-\delta}t^m\cdot(1-t)^{k-1}\cdot dt$$

$$-\sum_{m=0}^{[\alpha]}(-1)^m\cdot\binom{-\alpha}{m}\cdot\frac{\alpha+m}{\alpha-m}\cdot(1-\delta)^{m-\alpha}+\delta^{-\alpha}\cdot\left((1-\delta)^{-\alpha}-\sum_{m=0}^{[\alpha]}(-1)^m\cdot\binom{-\alpha}{m}\cdot\delta^m\right).$$

Recall that this expression does not depend on δ and that the terms of the double sum over m and k for which $k = 0$ are assumed to be zero. In addition, let us apply the Taylor expansion with Lagrange's form for the remainder to functions $(1-t)^{-\alpha}$ and $t^{-\alpha}$, in powers of t at the neighborhood of zero and in powers of $1-t$ at the neighborhood of one, respectively. Then we conclude the result that all the summands of the above sum do not have singularities as $\delta \downarrow 0$. Hence, the left-hand side of (1.2.27′) is equal to the limit of this sum as $\delta \downarrow 0$; let us now establish the value of this limit. It is easily seen that the limit of the first integral is equal to zero, the limit of the second summand coincides with the corresponding integral from zero to one, the limits of the third and the fourth summands are equal to the same expressions with the change of δ for zero, and the limit of the latter summand is equal to zero.

Thus, we obtain that for any positive non-integer α and for any positive $\delta < 1/2$, the left-hand side of (1.2.27′) is equal to

$$(1.2.28)\qquad \begin{aligned}&\int_0^1 ((1-t)^{-\alpha} - \sum_{m=0}^{[\alpha]} (-1)^m \cdot \binom{-\alpha}{m} \cdot t^m) \cdot d(-t^{-\alpha} + \sum_{m=0}^{[\alpha]} (-1)^m \cdot \binom{-\alpha}{m} \cdot (1-t)^m) \\ &- \sum_{m=0}^{[\alpha]} \sum_{k=1}^{[\alpha]} k \cdot (-1)^{m+k} \cdot \binom{-\alpha}{m} \binom{-\alpha}{k} \cdot \int_0^1 t^m \cdot (1-t)^{k-1} dt - \sum_{m=0}^{[\alpha]} (-1)^m \cdot \binom{-\alpha}{m} \cdot \frac{\alpha+m}{\alpha-m}.\end{aligned}$$

Now, note that for $\alpha \in (0,1)$, expression (1.2.28) turns into

$$(1.2.28') \qquad \int_0^1 ((1-t)^{-\alpha} - 1) \cdot d(-t^{-\alpha}).$$

Let us demonstrate that for these values of α, expression (1.2.28′) is equal to $-\Gamma(1-\alpha)^2 / \Gamma(1-2\cdot\alpha)$. To this end, we apply the well-known relationship between gamma- and beta- functions, which implies that for any real $u > 0$ and $v > 0$,

$$(1.2.29) \qquad B(u,v) = \int_0^1 t^{u-1} \cdot (1-t)^{v-1} \cdot dt = \frac{\Gamma(u) \cdot \Gamma(v)}{\Gamma(u+v)}.$$

Obviously, (1.2.29) yields that

$$(1.2.30) \qquad \int_0^1 t^{u-1} \cdot ((1-t)^{v-1} - 1) \cdot dt = B(u,v) - 1/u.$$

Note that the expressions on both sides of (1.2.30) admit analytic continuation onto $\{u \in \mathbf{C} : Re\, u > -1,\ u \neq 0\}$. Hence, by the uniqueness theorem on the analytic continuation, equality (1.2.30) holds also for $u \in \mathbf{C} \setminus \{0\}$, such that $Re\, u > -1$. Now, let us set $u := -\alpha$, and $v := 1 - \alpha$ in (1.2.30), recall that $\alpha \in (0,1)$ still. Then it follows from (1.2.30) that

$$\int_0^1 t^{-\alpha-1} \cdot ((1-t)^{-\alpha} - 1) \cdot dt = B(-\alpha, 1-\alpha) + 1/\alpha.$$

But the latter equality, in turn, is equivalent to the following one:

$$\int_0^1 ((1-t)^{-\alpha}-1)\cdot d(-t^{-\alpha}) - 1 = \alpha\cdot B(-\alpha,1-\alpha).$$

Applying (1.2.29) and the well-known properties of the gamma-function we get that the right-hand side of this equality (and hence the left-hand side as well) are equal to $-\Gamma(1-\alpha)^2/\Gamma(1-2\alpha)$.

Now, let us show that expression (1.2.28) is equal to $-\Gamma(1-\alpha)^2/\Gamma(1-2\alpha)$ for $\alpha \in (1,2)$ as well (Recall that for $\alpha = 1/2$ and $3/2$ this expression is assumed to be zero.) For this, we use the following equality, already established for $\alpha \in (0,1)$:

$$\int_0^1 ((1-t)^{-\alpha}-1)\cdot d(-t^{-\alpha}) - 1 = \alpha\cdot B(-\alpha,1-\alpha). \tag{1.2.31}$$

It is not difficult to show that for $\alpha \in (0,1)$, the left-hand side of (1.2.31) is equal to

$$\int_0^1 ((1-t)^{-\alpha}-1-\alpha t)\cdot d(-t^{-\alpha}+1+\alpha(1-t)) - 1 - \alpha\cdot\frac{\alpha+1}{\alpha-1} - \alpha\cdot(1+\alpha/2). \tag{1.2.32}$$

Obviously, expression (1.2.32) as well as the function $-\Gamma(1-\alpha)^2/\Gamma(1-2\alpha)$ admit analytic continuations from the interval $\alpha \in (0,1)$ onto $\{\alpha \in \mathbf{C} : Re\,\alpha \in (0,2)\} \setminus \{1\}$. Hence, the uniqueness theorem on the analytic continuation implies that expression (1.2.32) is equal to $-\Gamma(1-\alpha)^2/\Gamma(1-2\alpha)$ also for $\{\alpha \in \mathbf{C} : Re\,\alpha \in (0,2)\} \setminus \{1\}$. It only remains to note that for real $\alpha \in (1,2)$, expression (1.2.32) coincides with (1.2.28), which yields the required relationship.

Applying successively this scheme, we come to the conclusion that for any positive non-integer α, expression (1.2.28) coincides with the analytic continuation of $-\Gamma(1-\alpha)^2/\Gamma(1-2\alpha)$, which proves relationship (1.2.27′).

Now, let us establish the validity of (1.2.27). First, assume that $\alpha \in (0,1)$. Making the change of variables $u = (1-t)^{-1}$ we get that the left-hand side of (1.2.27) is equal to

$$\int_0^1 (u^\alpha - 1)\cdot d(|\frac{u-1}{u}|^{-\alpha}).$$

Integrating by parts, we obtain that this integral is equal to

$$-\int_0^1 (\frac{u}{1-u})^{-\alpha}\cdot du^\alpha = -\alpha\cdot\int_0^1 u^{2\alpha-1}\cdot(1-u)^{-\alpha}\cdot du = -\alpha\cdot \mathrm{B}(2\alpha,1-\alpha).$$

Together with (1.2.29) this yields the validity of (1.2.27) for $0 < \alpha < 1$. The proof of the validity of (1.2.27) for any positive non-integer α is carried out following along the same lines as that of (1.2.27′). □

1.3 Asymptotic Expansions of the Probabilities of Large Deviations of S_n in the Case of Quite Asymmetric Constraints on the Asymptotic Behavior of the Tails.

In this section, the results of Section 1.2 are generalized to wider classes of distributions. Namely, in Theorems 1.3.1 and 1.3.1′ we assume the fulfilment of Condition (0.5) with an arbitrary positive α_1 (including also the case of an integer α_1) and Condition (0.10) with an arbitrary positive non-integer β (the inequality $\alpha_1 \geq \beta$ is obtained automatically). Theorem 1.3.1 contains the asymptotic expansion for $\mathbf{P}\{S_n > y\}$ related to the case $\beta \in (0,1) \cup (1,2)$; this theorem describes both cases $\beta \leq \alpha_1 < 2$, and $\alpha_1 \geq 2$. Theorem 1.3.1′ is related to the case of non-integer $\beta > 2$.

The other result of this section, Theorem 1.3.2, is valid under fulfilment of Condition (0.5) with an arbitrary positive α_1 and under even weaker constraints on the asymptotics of the left-hand tail of $F(x)$. Namely, we only assume that r.v.'s $\{X_n, n \geq 1\}$ possess the finite absolute moment of order $\beta \geq 2$ (this automatically yields that $\alpha_1 > \beta \geq 2$).

The proofs of Theorems 1.3.1, 1.3.1′, and 1.3.2 are basically carried out by the direct probabilistic method described in the Introduction and then applied in the proofs of Theorems 1.1.1 and 1.2.2. The proofs of all the theorems of this section are analogous to that of Theorem 1.2.2. However, they are simplified by eliminating the terms of the alternating sum (0.8) beginning with the second rather than the third. Moreover, all the terms of the asymptotic expansion of the order $n^2 \cdot y^{-2\cdot\alpha_1}$ are absorbed into the remainders. Therefore, we only formulate Theorems 1.3.1, 1.3.1′, and 1.3.2, dropping their proofs. Note that in order to prove Theorems 1.3.1′ and 1.3.2, Corollaries 1.2.1 and 1.2.2 should be used, respectively.

To illustrate the results of this section, we also consider Example 1.3.1 to Theorem 1.3.2 which is related to the case when only the finiteness of the second moment of X_i's is known, and expansion (0.5) is valid with the remainder, $o(x^{-r})$, equal to zero identically. In other words, we have that for all sufficiently large x,

$$\mathbf{P}\{X_i > x\} = \sum_{i=1}^{l} c_{\alpha_i} \cdot x^{-\alpha_i}.$$

Let us point out that the asymptotic expansion for $\mathbf{P}\{S_n > y\}$ given in this example implies the exactness (in the sense of dependence on n and y) of the estimate of the rate of convergence in the following result on large deviations for S_n obtained in Theorem 2 of Kim and A.V.Nagaev (1972): for any integer $n \geq 4$ and $y \geq 2(\alpha+2)/(\alpha-2) \vee \sqrt{n} \cdot \log n$,

(1.3.1) $$|\mathbf{P}\{S_n > y\} \,/\, (n \cdot \mathbf{P}\{X_1 > y\}) - 1| \leq K \cdot n \cdot y^{-2}.$$

Note that Kim and A.V.Nagaev have obtained estimate (1.3.1) under the assumption that r.v.'s $\{X_n, n \geq 1\}$ have finite variance and some (significantly weaker) constraints on the asymptotics of the right-hand tail of $F(x)$ than those posed in the formulation of our Theorem 1.3.2.

Now, let us present the main results of this section.

Theorem 1.3.1. Let Condition (0.5) be fulfilled, and suppose there exists a constant K such that for any real $x > 0$,

$$F(-x) + 1 - F(x) \leq K \cdot x^{-\beta}$$

for some $\beta \in (0,1) \cup (1,2)$ (this is Condition (0.10) from the Introduction); $\mathbf{E}X_i = 0$ in the case $\beta \in (1,2)$. Then for any integer $n \geq 1$ and for any real $y > 0$,

$$\mathbf{P}\{S_n > y\} = n \cdot \sum_{i=1}^{l} c_{\alpha_i} \cdot y^{-\alpha_i} + \mathrm{r}_1(n,y) + \mathrm{r}_2(n,y),$$

where the remainders $\mathrm{r}_1(\cdot,\cdot)$ and $\mathrm{r}_2(\cdot,\cdot)$ are such that

i) for any integer $n \geq 1$ and for any real $y > 0$,

$$|r_1(n,y)| \leq n \cdot \sup_{x \geq y/\beta} \left|1 - F(x) - \sum_{i=1}^{l} c_{\alpha_i} \cdot x^{-\alpha_i}\right|;$$

ii) there exist positive constants K_1 and K_2 such that for any integer $n \geq 1$ and for $y \geq K_1 \cdot n^{1/\beta}$,

$$|r_2(n,y)| \leq K_2 \cdot n^2 \cdot y^{-\alpha_1 - \beta}.$$

In the next Theorems 1.3.1′ and 1.3.2 we use the function $M(\cdot,\cdot)$ introduced in Theorem 1.2.2. Recall that for any real $\gamma \geq 2$ and for any positive κ, $M(\gamma,\kappa) = [\gamma + (\gamma - 2) / (2\kappa)]$.

Theorem 1.3.1′. Let Conditions (0.5) and (0.10) be fulfilled with an arbitrary positive α_1, and non-integer $\beta > 2$; $\mathbf{E}X_i = 0$, and $\mathbf{V}X_i = 1$. In the case $\beta > 3$ we also assume the fulfilment of Condition (C). Then for any integer $n \geq 1$, and for any real $y > 0$ and $\kappa > 0$,

$$\mathbf{P}\{S_n > y\} = n \cdot \sum_{i=1}^{l} c_{\alpha_i} \cdot y^{-\alpha_i} \cdot \left(1 + \sum_{m=2}^{M(\beta,\kappa)} (-1)^m \cdot \binom{-\alpha_i}{m} \cdot y^{-m} \cdot (n-1)^{m/2}\right.$$

$$\left. \cdot \int_{-\infty}^{\infty} v^m \cdot d\left\{\Phi(v) + \sum_{\nu=1}^{m \wedge [\beta]-2} \frac{Q_\nu(v)}{(n-1)^{\nu/2}}\right\}\right) + r_1(n,y) + r_2^{(\kappa)}(n,y),$$

where the remainder $r_1(\cdot,\cdot)$ is the same as in Theorem 1.3.1, and the remainder $r_2^{(\kappa)}(\cdot,\cdot)$ is such that there exists a function $K(\cdot)$ from $\mathbf{R}_+^1$ into $\mathbf{R}_+^1$, such that for any integer $n \geq 1$, for any real $\kappa > 0$, and for $y \geq K(\kappa) \cdot n^{1/2+\kappa}$,

$$|r_2^{(\kappa)}(n,y)| \leq K_1 \cdot n^2 \cdot y^{-\alpha_1 - \beta}.$$

Theorem 1.3.2. Let Condition (0.5) be fulfilled, $\mathbf{E}|X_i|^\beta < \infty$ for some $\beta \geq 2$; $\mathbf{E}X_i = 0$, and $\mathbf{V}X_i = 1$. In the cases $\beta = 3$ and $\beta > 3$ we also assume that either r.v.'s $\{X_n, n \geq 1\}$ are not lattice or Condition (C) is fulfilled, respectively. Then for any integer $n \geq 1$, and for any real $y > 0$ and $\kappa > 0$,

$$\mathbf{P}\{S_n > y\} = n \cdot \sum_{i=1}^{l} c_{\alpha_i} \cdot y^{-\alpha_i} \cdot (1 + \sum_{m=2}^{M(\beta,\kappa)} (-1)^m \cdot \binom{-\alpha_i}{m} \cdot y^{-m} \cdot (n-1)^{m/2}$$

(1.3.2)

$$\cdot \int_{-\infty}^{\infty} v^{m} \cdot d\{\Phi(v) + \sum_{v=1}^{m\wedge[\beta]-2} \frac{Q_v(v)}{(n-1)^{v/2}} \}) + r_1(n,y) + r_2^{(\kappa)}(n,y),$$

where the remainder $r_1(\cdot,\cdot)$ is the same as in Theorem 1.3.1, and the remainder $r_2^{(\kappa)}(\cdot,\cdot)$ is such, that there exists a function $K(\cdot,\cdot)$ from $\mathbf{R}_+^1 \otimes \mathbf{R}_+^1$ into $\mathbf{R}_+^1$, such that for any real $\varepsilon > 0$ and $\kappa > 0$, for any integer $n \geq 1$, and for $y \geq K(\varepsilon,\kappa) \cdot n^{1/2+\kappa}$,

$$|r_2^{(\kappa)}(n,y)| \leq \epsilon \cdot n^2 \cdot y^{-\alpha_1-\beta}.$$

Example 1.3.1. Let i.i.d.r.v.'s $\{X_n, n \geq 1\}$ have only finite second moment, $\mathbf{E}X_i = 0$, $\mathbf{V}X_i = 1$, and Condition (0.5) be valid with the remainder, $o(x^{-r})$, equal to zero identically for all sufficiently large x, i.e., there exists $X_0 > 0$ such that for $x \geq X_0$,

$$\text{(1.3.3)} \qquad \mathbf{P}\{X_i > x\} = \sum_{i=1}^{l} c_{\alpha_i} \cdot x^{-\alpha_i}.$$

Obviously, all the conditions of Theorem 1.3.2 are fulfilled, but in this particular case, the right-hand side of (1.3.2) takes on the simplest form.

In fact, the sum over m consists of only one summand, since for any positive κ, $M(2,\kappa) \equiv 2$, the sum over v drops, and the remainder $r_1(\cdot,\cdot)$ is equal to zero identically for $y \geq 3 \cdot X_0$. Hence, Theorem 1.3.2 and expansion (1.3.3) imply that for any integer $n \geq 1$ and for $y \geq 3 \cdot X_0$,

$$\text{(1.3.4)} \qquad \begin{aligned} \mathbf{P}\{S_n > y\} &= n \cdot \mathbf{P}\{X_1 > y\} + n \cdot (n-1) \cdot \sum_{i=1}^{l} c_{\alpha_i} \cdot \binom{-\alpha_i}{2} \cdot y^{-\alpha_i-2} \\ &+ r_2^{(\kappa)}(n,y), \end{aligned}$$

where the remainder $r_2^{(\kappa)}(\cdot,\cdot)$ is such that there exists a function $K(\cdot,\cdot)$ from $\mathbf{R}_+^1 \otimes \mathbf{R}_+^1$ into $\mathbf{R}_+^1$, such that for any real $\varepsilon > 0$ and $\kappa > 0$, for any integer $n \geq 1$, and for $y \geq K(\varepsilon,\kappa) \cdot n^{1/2+\kappa}$,

$$|r_2^{(\kappa)}(n,y)| \leq \epsilon \cdot n^2 \cdot y^{-\alpha_1-2}.$$

Now, let us add and subtract the expression

$$n\cdot(n-1)\cdot\sum_{i=1}^{l} c_{\alpha_i}\cdot\binom{-\alpha_1}{2}\cdot y^{-\alpha_i-2}$$

to and from the right-hand side of (1.3.4), respectively. Then (1.3.3) implies that

$$\mathbf{P}\{S_n > y\} = n\cdot\mathbf{P}\{X_1 > y\}\cdot\left(1+\binom{-\alpha_1}{2}\cdot(n-1)\cdot y^{-2}\right)$$

$$+\ n\cdot(n-1)\cdot\sum_{i=2}^{l} c_{\alpha_i}\cdot\left(\binom{-\alpha_i}{2}-\binom{-\alpha_1}{2}\right)\cdot y^{-\alpha_i-2} + r_2^{(\kappa)}(n,y).$$

It is obvious that for any integer $n \geq 1$, for any real $\varepsilon > 0$, and for $y \geq \varepsilon^{-1/(\alpha_2-\alpha_1)}$, the sum over i on the right-hand side of the above representation does not exceed

$$K\cdot n^2\cdot y^{-\alpha_2-2} \leq K\cdot\epsilon\cdot n^2\cdot y^{-\alpha_1-2}.$$

After some transformations we get that for any integer $n \geq 1$, for any real $\varepsilon > 0$ and $\kappa > 0$, and for $y \geq K(\varepsilon,\kappa)\cdot n^{1/2+\kappa} \vee 3\cdot X_0 \vee \varepsilon^{-1/(\alpha_2-\alpha_1)}$,

$$\left|\frac{\mathbf{P}\{S_n > y\}}{n\cdot\mathbf{P}\{X_1 > y\}} - 1 - \binom{-\alpha_1}{2}\cdot(n-1)\cdot y^{-2}\right| \leq K_1\cdot\epsilon\cdot n\cdot y^{-2}.$$

In particular, this estimate yields the exactness (in the sense of dependence on n and y) of estimate (1.3.1), since the right-hand side of (1.3.1) is of the same order as the first refining term,

$$\binom{-\alpha_1}{2}\cdot(n-1)\cdot y^{-2}.$$

Chapter 2

Asymptotic Expansions of the Probabilities of Large Deviations and Non-Uniform Estimates of Remainders in CLT.

2.1 The Case of Power Tails with Integer Index $\alpha_1 \geq 3$.

In this section, we first extend the Edgeworth expansion and derive a non-uniform estimate of the remainder for this expansion in the case of power tails with integer index $\alpha_1 \geq 3$, and thereupon apply this result for the construction of the asymptotic expansion for $\mathbf{P}\{S_n > y\}$ in the range of large deviations of S_n (see Theorems 2.1.1 and 2.1.2, respectively). Note that in order to obtain the non-uniform estimate of Theorem 2.1.1, the fulfilment of slightly stronger conditions compare to (0. 1) - (0. 1′) is required. Namely, we assume that

$$(2.1.1) \qquad 1 - F(x) = c_{\alpha_1} \cdot x^{-\alpha_1} + O(x^{-\alpha_1 - \kappa});$$

$$(2.1.1') \qquad F(-x) = d_{\alpha_1} \cdot x^{-\alpha_1} + O(x^{-\alpha_1 - \kappa})$$

as $x \to \infty$, where $\kappa \in (0,1)$ is an arbitrary fixed real. In this section, we deal with the case of integer $\alpha_1 \geq 3$, whereas in Section 2.2 the case of $\alpha_1 = 2$ is considered. It should be pointed out that the t-distribution with the number of degrees of freedom $\alpha_1 \geq 3$ satisfies Conditions (2.1.1) - (2.1.1′) with the same α_1 and

$$c_{\alpha_1} = d_{\alpha_1} = \frac{\Gamma((\alpha_1 + 1)/2) \cdot \alpha_1^{(\alpha_1 - 2)/2}}{\Gamma(\alpha_1/2) \cdot \sqrt{\pi}}.$$

Note that coefficients which possess some properties of standard moments will emerge in all the expansions of this chapter. It is convenient to introduce these coefficients in advance, by means of

$$\mathscr{E}_{\alpha_1} := \int_{-\infty}^{-1} x^{\alpha_1} \cdot d(F(x) - d_{\alpha_1} \cdot |x|^{-\alpha_1}) + \int_{-1}^{1} x^{\alpha_1} \cdot dF(x)$$

$$+ \int_{1}^{\infty} x^{\alpha_1} \cdot d(F(x) + c_{\alpha_1} \cdot x^{-\alpha_1}), \tag{2.1.2}$$

where $\alpha_1 \geq 2$ is an integer. Here, the first and the third integrals on the right-hand side of (2.1.2) are understood as improper Stieltjes integrals. It is clear that under fulfilment of (2.1.1) - (2.1.1′), the pseudomoment $\mathscr{E}_{\alpha_1}$ is finite, since both improper integrals on the right-hand side of (2.1.2) converge.

Hereinafter, we refer to $\mathscr{E}_{\alpha_1}$ as the $\alpha_1{}^{st}$ pseudomoment in Ibragimov's sense; a special case of this concept was considered by Ibragimov (1967). Let us point out that pseudomoments in Ibragimov's sense differ from pseudomoments in Cramér's sense that will be considered in Chapter 3. Moreover, pseudomoments in Ibragimov's sense are related to the case of integer values of α_1, whereas those in Cramér's sense are related to the case of non-integer values of α_1. Now, we introduce three auxiliary sequences $\{E_+(k), k \geq 1\}$, $\{E_-(k), k \geq 1\}$, and $\{J_k, k \geq 1\}$:

$$E_+(k) := \int_0^1 (e^{iz} - \sum_{m=0}^{k} \frac{(iz)^m}{m!}) \cdot d(-z^{-k}) + \int_1^{\infty} (e^{iz} - \sum_{m=0}^{k-1} \frac{(iz)^m}{m!}) \cdot d(-z^{-k});$$

$$E_-(k) := \int_0^1 (e^{-iz} - \sum_{m=0}^{k} \frac{(-iz)^m}{m!}) \cdot d(-z^{-k}) + \int_1^{\infty} (e^{-iz} - \sum_{m=0}^{k-1} \frac{(-iz)^m}{m!}) \cdot d(-z^{-k});$$

$$J_k := \int_0^{\infty} (\cos z - \sum_{m=0}^{(k-1)/2} (-1)^m \cdot \frac{z^{2m}}{(2m)!}) \cdot d(-z^{-k}) \qquad \text{for odd } k;$$

$$J_k := \int_0^{\infty} (\sin z - \sum_{m=0}^{(k-2)/2} (-1)^{m+1} \cdot \frac{z^{2m+1}}{(2m+1)!}) \cdot d(-z^{-k}) \qquad \text{for even } k.$$

Simple calculations reveal that

$$J_k = (E_+(k) + E_-(k))/2 \tag{2.1.3}$$

for odd k, and

(2.1.3′) $J_k = (E_+(k) - E_-(k))/(2i)$

for even k.

It is easily seen that $J_1 = -\pi/2$, and $J_{k+1} = (-1)^{k+1} \cdot J_k / k$, which in turn imply that for any integer $k \geq 1$,

(2.1.4) $J_k = (-1)^{k(k+1)/2} \cdot \pi/(2(k-1)!)$.

Let us also introduce the following three families of functions:

$$A_k(x) := \frac{1}{2\pi} \cdot \int_{-\infty}^{\infty} e^{-t^2/2 - itx} \cdot E_{\mathrm{sgn}t}(k) \cdot |t|^k \cdot dt;$$

$$B_k(x) := \frac{1}{2\pi} \cdot \int_{-\infty}^{\infty} e^{-t^2/2 - itx} \cdot E_{-\mathrm{sgn}t}(k) \cdot |t|^k \cdot dt;$$

$$D_k(x) := \frac{1}{2\pi} \cdot \int_{-\infty}^{\infty} e^{-t^2/2 - itx} \cdot (it)^k \cdot \log \frac{1}{|t|} \cdot dt.$$

It should be mentioned that the functions $A_k(\cdot)$, $B_k(\cdot)$, $D_k(\cdot)$ are Fourier's transforms of absolutely integrable functions and therefore are continuous, bounded, and tend to zero as $|x| \to \infty$. Naturally, the above listed properties are also preserved for their linear combination

(2.1.5) $$h_{\alpha_1}(x) := c_{\alpha_1} \cdot A_{\alpha_1}(x) + d_{\alpha_1} \cdot B_{\alpha_1}(x) + \frac{e_{\alpha_1} \cdot D_{\alpha_1}(x)}{(\alpha_1 - 1)!},$$

where $e_{\alpha_1} := c_{\alpha_1} + (-1)^{\alpha_1} \cdot d_{\alpha_1}$.

Now set $\mathcal{H}_{\alpha_1}(x) := \int_{-\infty}^{x} h_{\alpha_1}(v) \cdot dv;$

this integral is finite in view of the properties of $h_{\alpha_1}(\cdot)$. Moreover,

$$\mathcal{H}_{\alpha_1}(-\infty) = \mathcal{H}_{\alpha_1}(\infty) = 0.$$

For integer $\alpha_1 \geq 3$, set

$$\Delta_{n,\alpha_1}(x) := F_n(x) - \Phi(x) - \sum_{v=1}^{\alpha_1 - 2} \frac{Q_v(x)}{n^{v/2}} + \frac{e_{\alpha_1} \cdot \log n \cdot \Phi^{(\alpha_1)}(x)}{2(\alpha_1 - 1)! \cdot n^{(\alpha_1 - 2)/2}} - \frac{\mathcal{H}_{\alpha_1}(x)}{n^{(\alpha_1 - 2)/2}}.$$

Recall that $F_n(x) := \mathbf{P}\{S_n \leq x \cdot \sqrt{n}\}$; the functions $Q_v(\cdot)$ are expressed in terms of the moments of r.v. X_1 by the formulas which can be found in Petrov (1975a), Chapter VI, (1.13) with the exception that one should use the α_1^{st} pseudomoment in Ibragimov's sense $\mathscr{E}_{\alpha_1}$ instead of the α_1^{st} moment in the formula for function $Q_{\alpha_1-2}(\cdot)$.

The next result is analogous to Propositions 1.2.1 - 1.2.3 from Section 1.2:

Theorem 2.1.1. Let Conditions (2.1.1) - (2.1.1′) be fulfilled with an integer $\alpha_1 \geq 3$; $\mathbf{E}X_i = 0$, and $\mathbf{V}X_i = 1$. In addition, let us assume that either r.v. X_1 is not lattice if $\alpha_1 = 3$ or Condition (**C**) is fulfilled if $\alpha_1 \geq 4$. Then

i) for any integer $n \geq 1$ and for any real x,

$$(2.1.6) \qquad |\Delta_{n,\alpha_1}(x)| \leq \frac{\delta(n)}{n^{(\alpha_1-2)/2}\cdot(1+|x|)^{\alpha_1}}.$$

ii) Function $\mathcal{H}_{\alpha_1}(\cdot)$ has the following asymptotics on the tails:

$$\mathcal{H}_{\alpha_1}(\infty) - \mathcal{H}_{\alpha_1}(x) = -\mathcal{H}_{\alpha_1}(x) = c_{\alpha_1}\cdot x^{-\alpha_1} + O(x^{-\alpha_1-1}\cdot\log x);$$

$$\mathcal{H}_{\alpha_1}(-x) = d_{\alpha_1}\cdot x^{-\alpha_1} + O(x^{-\alpha_1-1}\cdot\log x)$$

as $x \to \infty$.

Proof of Theorem 2.1.1. First, let us establish the assertions of point (ii). This can be done by applying the stationary phase method for the derivation of the asymptotics of the tails of the functions $D_{\alpha_1}(\cdot)$, $A_{\alpha_1}(\cdot)$, and $B_{\alpha_1}(\cdot)$. In particular, the tail behavior of functions $D_{\alpha_1}(\cdot)$, $A_{\alpha_1}(\cdot)$, and $B_{\alpha_1}(\cdot)$ is obtained in Points (ii1) - (ii3), respectively.

(ii1) In order to express $D_{\alpha_1}(\cdot)$ in a more convenient form, let us make the change of variables v = - x in the integral from $-\infty$ to 0. We get that

$$(2.1.7) \qquad D_{\alpha_1}(x) = -\frac{i^{\alpha_1}}{2\pi}\int_0^\infty e^{-t^2/2}\cdot t^{\alpha_1}\cdot\log t\cdot\left(e^{-itx} + (-1)^{\alpha_1}\cdot e^{itx}\right)\cdot dt.$$

Set $$\mathrm{I}_{\alpha_1} := \int_0^\infty e^{-t^2/2+itx}\cdot t^{\alpha_1}\log t\cdot dt.$$

Now, applying (2.1.1) and the well-known properties of function e^{itx} one gets that

(2.1.8) $$D_{\alpha_1}(x) = i^{\alpha_1+1}\cdot\frac{Im\ I_{\alpha_1}(x)}{\pi} \qquad \text{for odd } \alpha_1;$$

(2.1.8′) $$D_{\alpha_1}(x) = -i^{\alpha_1}\cdot\frac{Re\ I_{\alpha_1}(x)}{\pi} \qquad \text{for even } \alpha_1.$$

In order to derive the asymptotics of $I_{\alpha_1}(\cdot)$, we apply Lemma 1.4 from Chapter 3 of Fedoryuk (1987). We obtain that

(2.1.9) $$Im\ I_{\alpha_1}(x) = \pm i^{\alpha_1+1}\cdot\Gamma(\alpha_1+1)\cdot\pi\cdot x^{-\alpha_1-1}/2 + O(|x^{-\alpha_1-2}|\cdot\log|x|)$$

as $x \to \pm\infty$ for odd α_1, and

(2.1.9′) $$Re\ I_{\alpha_1}(x) = i^{\alpha_1+2}\cdot\Gamma(\alpha_1+1)\cdot\pi\cdot|x|^{-\alpha_1-1}/2 + O(|x^{-\alpha_1-2}|\cdot\log|x|)$$

as $x \to \pm\infty$ for even α_1. A subsequent combination of (2.1.8) - (2.1.8′), and (2.1.9) - (2.1.9′) implies that

(2.1.10) $$D_{\alpha_1}(x) = \begin{cases} \pm\alpha_1!\cdot x^{-\alpha_1}/2 + O(|x|^{-\alpha_1-2}\cdot\log|x|) & \text{for odd } \alpha_1; \\ \alpha_1!\cdot|x|^{-\alpha_1-1}/2 + O(|x|^{-\alpha_1-2}\cdot\log|x|) & \text{for even } \alpha_1 \end{cases}$$

as $x \to \pm\infty$.

(ii2) In order to find the tail behavior of function $A_{\alpha_1}(\cdot)$ let us make the same change of variables $v = -x$ in the integral from $-\infty$ to 0 as in (ii1). We get that

$$A_{\alpha_1}(x) = \frac{1}{2\pi}\int_0^\infty e^{-t^2/2}\cdot t^{\alpha_1}\cdot\left(E_+(\alpha_1)\cdot e^{-itx} + E_-(\alpha_1)\cdot e^{itx}\right)\cdot dt.$$

Set

(2.1.11) $$I_+(\alpha_1,x) := \int_0^\infty e^{itx-t^2/2}\cdot t^{\alpha_1}\cdot dt;$$

(2.1.11′) $$I_-(\alpha_1,x) := \int_0^\infty e^{-itx-t^2/2}\cdot t^{\alpha_1}\cdot dt.$$

It is easily seen that

(2.1.12) $$Re\ I_+(\alpha_1, x) = Re\ I_-(\alpha_1, x),$$

and

(2.1.12′) $Im\, I_+(\alpha_1, x) = -\, Im\, I_-(\alpha_1, x)$.

Using the above introduced notation, one can represent $A_\alpha(x)$ in the following form:

(2.1.13) $$A_{\alpha_1}(x) = \frac{E_+(\alpha_1)\cdot I_-(\alpha_1,x) + E_-(\alpha_1)\cdot I_+(\alpha_1,x)}{2\pi}.$$

Keeping in mind (2.1.11) - (2.1.11′) and integrating by parts, one obtains that

$$I_+(\alpha_1,x) = \alpha_1!\cdot(-ix)^{-\alpha_1-1} + O(|x|^{-\alpha_1-2})$$

as $x \to \pm\infty$. Hence,

(2.1.14) $$I_+(\alpha_1,x) = Re\, I_+(\alpha_1,x) + O(|x|^{-\alpha_1-2})$$

for odd α_1, and

(2.1.14′) $$I_+(\alpha_1,x) = i\cdot Im\, I_+(\alpha_1,x) + O(|x|^{-\alpha_1-2})$$

for even α_1 as $x \to \pm\infty$. In addition,

(2.1.15) $$Re\, I_+(\alpha_1, x) = \alpha_1!\cdot(-ix)^{-\alpha_1-1} + O(|x|^{-\alpha_1-2})$$

for odd α_1 , and

(2.1.15′) $$i\cdot Im\, I_+(\alpha_1, x) = \alpha_1!\cdot(-ix)^{-\alpha_1-1} + O(|x|^{-\alpha_1-2})$$

for even α_1 as $x \to \pm\infty$.

A combination of (2.1.12) - (2.1.15) and (2.1.12′) - (2.1.15′) yields that

$$A_{\alpha_1}(x) = \begin{cases} \dfrac{\alpha_1!\big(E_+(\alpha_1) + E_-(\alpha_1)\big)\cdot(-ix)^{-\alpha_1-1}}{2\pi} + O(|x|^{-\alpha_1-2}) & \text{for odd } \alpha_1; \\[2ex] \dfrac{\alpha_1!\big(E_-(\alpha_1) - E_+(\alpha_1)\big)\cdot(-ix)^{-\alpha_1-1}}{2\pi} + O(|x|^{-\alpha_1-2}) & \text{for even } \alpha_1 \end{cases}$$

as $x \to +\infty$. Combining the above representation for $A_\alpha(\cdot)$ with (2.1.3), (2.1.3′), and (2.1.4) we get that

(2.1.16) $$A_{\alpha_1}(x) = \frac{\alpha_1\cdot x^{-\alpha_1-1}}{2} + O(|x|^{-\alpha_1-2})$$

as $x \to \pm\infty$.

(ii3) By analogy with (2.1.16) we obtain that

$$(2.1.17)\qquad B_{\alpha_1}(x) = \frac{\alpha_1\cdot(-x)^{-\alpha_1-1}}{2} + O(|x|^{-\alpha_1-2})$$

as $x \to \pm\infty$. To complete the proof of the assertions of point (ii), it only remains to combine (2.1.5), (2.1.10), and (2.1.16) - (2.1.17). Indeed, one gets that

$$(2.1.18)\qquad -\mathcal{H}_{\alpha_1}(x) := \int_x^{\infty} h_{\alpha_1}(v)\cdot dv = c_{\alpha_1}\cdot x^{-\alpha_1} + O(x^{-\alpha_1-1}\cdot\log x),$$

and

$$(2.1.18')\qquad \mathcal{H}_{\alpha_1}(-x) := \int_{-\infty}^{-x} h_{\alpha_1}(v)\cdot dv = d_{\alpha_1}\cdot x^{-\alpha_1} + O(x^{-\alpha_1-1}\cdot\log x)$$

as $x \to \infty$. □

Secondly, in order to derive (2.1.6), the exact (in the sense of dependence on n) uniform estimate for $|\Delta_{n,\alpha_1}(x)|$ should be established:

$$(2.1.19)\qquad |\Delta_{n,\alpha_1}(x)| \le \frac{K}{n^{(\alpha_1+\kappa-2)/2}}.$$

The proof of (2.1.19) is straightforward and relies on the following representation of the characteristic function $f(t)$ of r.v. X_1 near zero:

$$(2.1.20)\qquad \begin{aligned} f(t) &= 1 + \sum_{k=2}^{\alpha_1} \frac{\mathcal{E}_k(it)^k}{k!} + \frac{e_{\alpha_1}}{(\alpha_1-1)!}\cdot(it)^{\alpha_1}\cdot\log\frac{1}{|t|} \\ &+ |t|^{\alpha_1}\cdot\left(c_{\alpha_1}\cdot E_{\operatorname{sgn} t}(\alpha_1) + d_{\alpha_1}\cdot E_{-\operatorname{sgn} t}(\alpha_1)\right) + O(|t|^{\alpha_1+\kappa}) \end{aligned}$$

as $|t| \to 0$. Note also that estimate (2.1.19) yields the fulfilment of estimate (2.1.6) both in the range of normal deviations and for $|x| \to \infty$, such that $x = o(n^{\kappa/2\alpha_1})$. Hence, it only remains to consider the range of deviations $|x| \ge K\cdot n^{\kappa/2\alpha_1}$. But the assertion of point (i) in this range of large deviations can be obtained by the use of (2.1.18) - (2.1.18′) and Corollary 1.1.2 of Section 1.1 with $\beta = \alpha_1$. In particular, one gets that

$$\left|\mathbf{P}\{S_n < -y\} - \frac{\mathcal{H}_{\alpha_1}(-y/\sqrt{n})}{n^{(\alpha_1-2)/2}}\right| \vee \left|\mathbf{P}\{S_n > y\} + \frac{\mathcal{H}_{\alpha_1}(y/\sqrt{n})}{n^{(\alpha_1-2)/2}}\right|$$

$$(2.1.21) \quad \le K_1 \cdot ny^{-\alpha_1} \cdot \left(\left(\frac{y}{\sqrt{n}}\right)^{-\alpha_1/(1+\alpha_1)} + \exp\left\{-K_2 \cdot \left(\frac{y^2}{\sqrt{n}}\right)^{1/(1+\alpha_1)}\right\}\right)$$

$$+ \; K_3 \cdot ny^{-\alpha_1-\kappa} + K_4 \cdot ny^{-\alpha_1}\left(\frac{\sqrt{n}}{y}\right) \cdot \log\left(\frac{y}{\sqrt{n}}\right).$$

It only remains to replace $\mathbf{P}\{S_n < -y\}$ by $F_n(-x)$; and $\mathbf{P}\{S_n > y\}$ by $1 - F_n(x)$ in the expression on the left-hand side of (2.1.21), and use the fact that the Laplace function $\Phi(\cdot)$ exponentially decays on the tails (Recall that $x = y/\sqrt{n}$.) Therefore, we can rewrite (2.1.21) as follows:

$$\left|F_n(x) - \Phi(x) - \frac{\mathcal{H}_{\alpha_1}(x)}{n^{(\alpha_1-2)/2}}\right|$$

$$\le K_1 n^{\frac{-(\alpha_1-2)}{2}} |x|^{-\alpha_1} \left\{ |x|^{\frac{-\alpha_1}{1+\alpha_1}} + \exp\{-K_2 |x|^{\frac{2}{1+\alpha_1}}\} + |x|^{-\kappa} + \frac{\log|x|}{|x|} \right\}$$

as $n \to \infty$ with $|x| \ge (K \cdot \log n)^{1/2}$. A combination of the latter inequality with the exponential decay of functions $Q_\nu(\cdot)$ and $\Phi^{(\alpha_1)}(\cdot)$ implies (2.1.6). □

It is interesting to note that one can employ (2.1.6) in order to obtain the following result on the asymptotic behavior of the probabilities of large deviations of S_n in the full range:

Corollary 2.1.1. Under the conditions of Theorem 2.1.1,

$$\mathbf{P}\{S_n > y\} = 1 - \Phi\left(\frac{y}{\sqrt{n}}\right) - \sum_{\nu=1}^{\alpha_1-2} \frac{Q_\nu(y/\sqrt{n})}{n^{\nu/2}} + \frac{e_{\alpha_1} \cdot \log n \cdot \Phi^{(\alpha_1)}(y/\sqrt{n})}{2(\alpha_1-1)! \cdot n^{(\alpha_1-2)/2}}$$

$$+ \; n \cdot c_{\alpha_1} \cdot y^{-\alpha_1} + o(n \cdot y^{-\alpha_1})$$

as $n \to \infty$, $y/\sqrt{n} \to \infty$.

Remark 2.1.1. It is clear that Corollary 2.1.1 refines relationship (0.3) for the special case of fulfilment of Conditions (2.1.1) - (2.1.1′) and in the range of large deviations $y/\sqrt{n} \to \infty$, $y \le (K \cdot n \cdot \log n)^{1/2}$.

Now, let us proceed with the derivation of the asymptotic expansion for the probabilities of large deviations of S_n. The next result is proved by the use of Theorem 2.1.1 and Corollary 1.1.2 from Section 1.1:

Theorem 2.1.2. Let Conditions (0.5) and (2.1.1′) be fulfilled with an integer $\alpha_1 \geq 3$; $\mathbf{E}X_i = 0$, and $\mathbf{V}X_i = 1$. In addition, let us assume that either r.v. X_1 is not lattice if $\alpha_1 = 3$ or Condition (**C**) is fulfilled if $\alpha_1 \geq 4$.

Then for any integer $n \geq 2$ and for any real $y > 0$,

$$\mathbf{P}\{S_n > y\} = n \cdot \sum_{i=1}^{l} c_{\alpha_i} \cdot y^{-\alpha_i} \cdot \left(1 + \sum_{m=2}^{M(\alpha_1,\theta)} (-1)^m \cdot \binom{-\alpha_i}{m} \cdot y^{-m} \cdot (n-1)^{m/2} \cdot L(m,n-1)\right)$$
$$+ \frac{n(n-1)}{2} \cdot \log\log(e \cdot (n-1)) \cdot c_{\alpha_1} \cdot \binom{-\alpha_1}{\alpha_1} \cdot \alpha_1 \cdot \varepsilon_{\alpha_1} \cdot y^{-2\alpha_1} + r_1(n,y) + r_2^{(\theta)}(n,y),$$

where

$L(m,0) := 1$ for any integer $m \geq 2$;

$L(2,n\text{-}1) := 1$ for any integer $n \geq 2$;

$$L(m,n-1) := \int_{-\infty}^{\infty} v^m \cdot d\left\{\Phi(v) + \sum_{v=1}^{\alpha_1-2} \frac{Q_v(v)}{(n-1)^{v/2}}\right\}$$

if $n \geq 2$ for any $1 < m \leq \alpha_1 - 1$, and

$$L(m,n-1) := \int_{-\infty}^{\infty} v^m \cdot d\left\{\Phi(v) + \sum_{v=1}^{\alpha_1-2} \frac{Q_v(v)}{(n-1)^{v/2}} - \frac{\varepsilon_{\alpha_1} \cdot \log n \cdot \Phi^{(\alpha_1)}(v)}{2(\alpha_1-1)! \cdot (n-1)^{(\alpha_1-2)/2}}\right\}$$

if $n \geq 2$ for any $m \geq \alpha_1$; $M(\alpha_1, \theta) := \alpha_1 + 1 + [(\alpha_1 - 2) / (2\theta)]$; the remainder $r_1(\cdot,\cdot)$ is such that for any integer $n \geq 1$ and for any real $y > 0$,

$$|r_1(n,y)| \leq n \cdot \sup_{x \geq y/3} \left|1 - F(x) - \sum_{i=1}^{l} c_{\alpha_i} \cdot x^{-\alpha_i}\right|,$$

and the remainder $r_2^{(\theta)}(\cdot,\cdot)$ is such that there exists a function $\mathbf{K}(\cdot)$ from $\mathbf{R}_+^1$ into $\mathbf{R}_+^1$, such that for any real $\theta > 0$, for any integer $n \geq 1$, and for $y \geq K(\theta) \cdot n^{1/2+\theta}$,

$$|r_2^{(\theta)}(n,y)| \leq K_1 \cdot (n \cdot y^{-\alpha_1})^2.$$

Remark 2.1.2. Note that in the preliminary formulation of this result (cf. Theorem 2.1

of Vinogradov (1990)) the term

$$\frac{n(n-1)}{2}\cdot\log\log(e\cdot(n-1))\cdot c_{\alpha_1}\cdot\binom{-\alpha_1}{\alpha_1}\cdot\alpha_1\cdot e_{\alpha_1}\cdot y^{-2\alpha_1}$$

in the asymptotic expansion for $\mathbf{P}\{S_n > y\}$ was missed.

Proof of Theorem 2.1.2 follows along the same lines as that of Theorem 1.2.2 of Section 1.2 and therefore is omitted. Note that instead of the non-uniform estimate of Corollary 1.2.1 we should use the non-uniform estimate of Theorem 2.1.1 and also the uniform estimate (2.1.19). □

2.2 The Case of Power Tails with Index $\alpha_1 = 2$.

In this section, we first construct an asymptotic expansion in the limit theorem on non-normal convergence to the normal law along with the non-uniform estimate of the remainder for the random sequence

$$\varsigma_n := S_n\,/\,((c_2 + d_2)\cdot n\cdot\log n)^{1/2}$$

under fulfilment of Conditions (2.1.1) - (2.1.1′) with $\alpha_1 = 2$ (cf. Theorem 2.2.1). It is of interest to note that the t-distribution with two degrees of freedom satisfies the just mentioned conditions with $\alpha_1 = 2$, and $c_2 = d_2 = 1/2$. The result of Theorem 2.2.1 can have its own value as due to novelty of some refining terms of the asymptotic expansion as well as due to the fact that it yields a more precise representation (compare to relationship (0.2′)) for the probabilities of large deviations of S_n (cf. Corollary 2.2.1 below). The conclusive result of this section, Theorem 2.2.2, contains the asymptotic expansion for $\mathbf{P}\{S_n > y\}$. It is interesting to note that one term of that expansion contains a triple logarithm.

Set $$\mathcal{H}_2(x) := \int_{-\infty}^{x} dv\cdot\left(\int_{-\infty}^{\infty}\frac{e^{-\frac{t^2}{2}-itv}}{2\pi}\cdot\left((it)^2\cdot\log\frac{1}{|t|}+\frac{c_2\cdot E_{\operatorname{sgn} t}+d_2\cdot E_{-\operatorname{sgn} t}}{e_2}\right)dt\right),$$

and $F_n(x) := \mathbf{P}\{\varsigma_n \le x\}$. Then the following result is valid:

Theorem 2.2.1. Let Conditions (2.1.1) - (2.1.1′) be fulfilled with $\alpha_1 = 2$,

$e_2 := c_2 + d_2 > 0$, $\mathbf{E}X_i = 0$, and r.v.'s $\{X_n, n \geq 1\}$ be not lattice. Then

(i) for any integer $n \geq 2$ and for any real x,

$$(2.2.1) \qquad \left| F_n(x) - \Phi(x) - \frac{\log\log n + \log e_2 + \frac{\mathcal{E}_2}{e_2}}{2 \cdot \log n} \cdot \Phi''(x) - \frac{\mathcal{H}_2(x)}{\log n} \right| \leq \frac{\delta(n)}{(1+|x|)^2 \cdot \log n},$$

where $\delta(n) \to 0$ as $n \to \infty$.

(ii) Function $\mathcal{H}_2(\cdot)$ has the following asymptotics on the tails:

$$(2.2.2) \qquad -\mathcal{H}_2(x) = \frac{c_2}{e_2} \cdot x^{-2} + O(x^{-3} \cdot \log x);$$

$$(2.2.2') \qquad \mathcal{H}_2(-x) = \frac{d_2}{e_2} \cdot x^{-2} + O(x^{-3} \cdot \log x)$$

as $x \to \infty$.

Remark 2.2.1. It is interesting to compare (2.2.1) with uniform estimates that can be derived from Theorems 2 and 5 of Hall (1983) (see also Theorem 4.12 of Hall (1982)). In particular, Theorem 4.12 of Hall (1982) implies that under the conditions of our Theorem 2.2.1, the following relationship holds:

$$\mathbf{P}\{S_n \leq C_n \cdot x + D_n\} - \Phi(x) - \left(\frac{\log\log n}{2 \cdot \log n} + \frac{\psi(x,n)}{2 \cdot \log n} \right) \cdot \Phi''(x) = o\left(\frac{1}{\log n}\right)$$

as $n \to \infty$ uniformly in $x \in \mathbf{R}^1$, where

$$C_n := \sup\left\{a: a^{-2} \cdot \mathbf{E}(X_i^2 \cdot \mathrm{I}_{(|X_i| \leq a)}) \geq \frac{1}{n}\right\} \sim \sqrt{(c_2 + d_2) \cdot n \cdot \log n}$$

as $n \to \infty$,

$$D_n := n \cdot \mathbf{E}(X_i^2 \cdot \mathrm{I}_{(|X_i| \leq C_n)}) \sim (c_2 - d_2) \cdot n \cdot \log n$$

as $n \to \infty$, $\psi(\cdot,\cdot)$ is a certain uniformly bounded real-valued function from $\mathbf{R}^1 \times \mathbf{N}$, and I_A stands for the indicator of set A. On the other hand, the asymptotic expansion of Theorem 5 of Hall (1983) contains a refining term

$$2n \cdot \mathbf{P}\{|X_n| > C_n\} \cdot \omega(\Phi(x)),$$

where the operator $\omega(\cdot)$ is defined in Höglund (1970) formula (4). In particular,

$$\omega(\Phi(x)) = \int_{-\infty}^{+\infty} \frac{\Phi(x-y)-\Phi(x)+y\Phi'(x)-\tau^2(y)\Phi''(x)/2}{y^3} \cdot \Omega(dy),$$

where $\tau(x) := x$ if $|x| \leq 1$ and $\tau(x) := 0$ if $|x| > 1$, and function $\Omega(\cdot)$ is defined as follows:

$$\Omega(x) = \begin{cases} c_2 \cdot x/e_2 & \text{if } x > 0, \\ d_2 \cdot x/e_2 & \text{if } x < 0. \end{cases}$$

Let us point out that this term corresponds to the term $\mathcal{H}_2(x)/\log n$ in (2.2.1). However, note that the just mentioned results by Hall provide uniform estimates of remainders, whereas our estimate (2.2.1) is non-uniform.

Proof of Theorem 2.2.1. Proof of (ii) is carried out by the stationary phase method. It is similar to that of Theorem 2.1.1.ii and therefore is omitted.

Now, in order to establish (i) we need to derive the following uniform estimate for the expression on the left-hand side of (2.2.1):

Proposition 2. 2. 1. Let all the conditions of Theorem 2.2.1 be fulfilled. Then there exists a positive constant K such that for any integer $n \geq 2$ and for any real x, the left-hand side of (2.2.1) does not exceed

(2.2.3) $$K / (n^{\kappa/2} \cdot (1 + \log n)^{1+\kappa/2}).$$

Proof of Proposition 2.2.1 is straightforward. It relies on the Smoothing Inequality and the following representation of the characteristic function $f(t)$ of r.v. X_1 as $t \to 0$ (compare to Theorem 3 of Pitman (1968)):

$$f(t) = 1 + \mathcal{E}_2 \cdot (it)^2 / 2 + e_2 \cdot (it)^2 \cdot \log 1/|t|$$

(2.2.4) $$+ |t|^2 \cdot (c_2 \cdot E_{\operatorname{sgn} t}(2) + d_2 \cdot E_{-\operatorname{sgn} t}(2)) + O(|t|^{2+\kappa}).$$

Then point (i) of Theorem 2.2.1 follows from the uniform estimate (2.2.3), relationships (2.2.2) - (2.2.2′) of point (ii), Corollary 1.1.3 of Section 1.1, and the following bound (that can be viewed as the left-hand analog of Corollary 1.1.3):

$$|\mathbf{P}\{S_n < -y\} - n \cdot d_2 \cdot y^{-2}| \le K \cdot n \cdot y^{-2} \cdot n^{1/3} \cdot y^{-2/3} + n \cdot \sup_{x \le -y/3} |F(x) - d_2 \cdot x^{-2}|.$$

In fact, (2.2.3) yields (2.2.1) in the range of deviations $|x| \le \delta_1(n) \cdot (n \cdot \log n)^{\kappa/4}$, where $\delta_1(n) \to 0$ as $n \to \infty$. On the other hand, replacing y by $x \cdot (\mathfrak{e}_2 \cdot n \cdot \log n)^{1/2}$ in Corollary 1.1.3 and its left-hand analog we obtain upper bounds for

$$\left|1 - F_n(x) - \frac{c_2 \cdot x^{-2}}{\mathfrak{e}_2 \cdot \log n}\right|$$

and for

$$\left|F_n(-x) - \frac{d_2 \cdot x^{-2}}{\mathfrak{e}_2 \cdot \log n}\right|$$

in the range of large deviations $x \ge ((K/\mathfrak{e}_2) \cdot \log\log(e \cdot n))^{1/2}$. A subsequent application of relationships (2.2.2) - (2.2.2′) yields the required estimates for

$$|1 - F_n(x) + \mathcal{H}_2(x) / \log n|$$

in the range of deviations $x \ge ((K/\mathfrak{e}_2) \cdot \log\log(e \cdot n))^{1/2}$ and for

$$|F_n(x) - \mathcal{H}_2(x) / \log n|$$

in the range of deviations $x \le -((K/\mathfrak{e}_2) \cdot \log\log(e \cdot n))^{1/2}$. It only remains to combine these estimates with the exponential decay of the tails of Φ and Φ''. □

Corollary 2.2.1. Let all the conditions of Theorem 2.2.1 be fulfilled. Then

$$\mathbf{P}\{S_n > y\} = 1 - \Phi\left(\frac{y}{\sqrt{\mathfrak{e}_2 \cdot n \cdot \log n}}\right) - \frac{\log\log n + \log \mathfrak{e}_2 + \mathcal{E}_2/\mathfrak{e}_2}{2 \cdot \log n} \tag{2.2.5}$$

$$\cdot\ \Phi''\left(\frac{y}{\sqrt{\mathfrak{e}_2 \cdot n \cdot \log n}}\right) + n \cdot c_2 \cdot y^{-2} + o(n \cdot y^{-2}).$$

Remark 2.2.2. Note that for a special case of fulfilment of Conditions (2.1.1) - (2.1.1′) with index $\alpha_1 = 2$, our representation (2.2.5) for $\mathbf{P}\{S_n > y\}$ is more precise, compare to relationship (0.2′), in the range of deviations $y / (n \cdot \log n)^{1/2} \to \infty$, $y \le const \cdot (n \cdot \log n \cdot \log\log n)^{1/2}$. This is on account of the fact that the right-hand side of (2.2.5) contains an asymptotic expansion for $\mathbf{P}\{S_n > y\}$ in this range of deviations rather than the exact asymptotics for $\mathbf{P}\{S_n > y\}$, given by relationship (0.2′).

On the other hand, it is easily seen that in the range of deviations $y / (n \log n \cdot \log\log n)^{1/2} \to \infty$ as $n \to \infty$, Corollary 2.2.1 provides only the asymptotics of $\mathbf{P}\{S_n > y\}$ up to equivalence (compare to (0.2′)). Nonetheless, we cannot guarantee their accuracy better than up to $const \cdot n \cdot y^{-2-\kappa}$ (that is not much different from $o(n \cdot y^{-2})$) at least because that is the case even for $n = 1$ (cf. (2.1.1)). Therefore, in order to get a more detailed information on the asymptotic behavior of the probabilities of right-hand deviations of S_n in this range of deviations, we should also require the fulfilment of Condition (0.5) with $\alpha_l = 2$:

Theorem 2.2.2. Let Conditions (0.5) and (2.1.1′) be fulfilled with $\alpha_l = 2$, $\varepsilon_2 > 0$, $EX_i = 0$, and r.v.'s $\{X_n, n \geq 1\}$ be not lattice. Then for any integer $n \geq 2$ and for any real $y > 0$,

$$\mathbf{P}\{S_n > y\} = n \cdot \sum_{i=1}^{l} c_{\alpha_i} \cdot y^{-\alpha_i} + 6 \cdot n \cdot (n-1) \cdot \log(n-1) \cdot c_2 \cdot \varepsilon_2 \cdot y^{-4} + 9 \cdot n$$

$$\cdot (n-1) \cdot \log\log(e \cdot (n-1)) \cdot c_2 \cdot \varepsilon_2 \cdot y^{-4} + 6 \cdot n \cdot (n-1) \cdot \log\log\log\left(e^e \cdot (n-1)\right)$$

$$\cdot c_2 \cdot \varepsilon_2 \cdot y^{-4} + r_1(n,y) + r_2^{(\theta)}(n,y),$$

where the remainder $r_1(\cdot,\cdot)$ is such that for any integer $n \geq 2$ and for any real $y > 0$,

$$|r_1(n,y)| \leq n \cdot \sup_{x \geq y/3} \left|1 - F(x) - \sum_{i=1}^{l} c_{\alpha_i} \cdot x^{-\alpha_i}\right|,$$

and the remainder $r_2^{(\theta)}(\cdot,\cdot)$ is such that there exists a function $K(\cdot)$ from $\mathbf{R}_+^1$ into $\mathbf{R}_+^1$, such that for any real $\theta > 0$, for any integer $n \geq 2$, and for $y \geq K(\theta) \cdot n^{1/2+\theta}$,

$$|r_2^{(\theta)}(n,y)| \leq K_1 \cdot n^2 \cdot y^{-4},$$

Proof of Theorem 2.2.2 follows along the same lines as that of Theorem 1.2.2 and therefore is omitted. Note that instead of the non-uniform estimate of Corollary 1.2.1 we should use the non-uniform estimate of Theorem 2.2.1 and also the uniform estimate (2.2.3). □

Chapter 3

Asymptotic Expansions Taking into Account the Cases when the Number of Summands Comparable with the Sum Does not Exceed a Fixed Integer.

In Chapters 1 and 2 we have constructed the asymptotic expansions of the probabilities of large deviations of S_n under the fulfilment of Condition (0.5) and certain supplementary constraints on the left-hand tail of $F(x)$. Note that even in the most precise expansions of those chapters, namely, Theorems 1.2.1 and 1.2.2 of Section 1.2, only the two initial terms of the alternating sum (0.8) are taken into account.

Recall that the nature of large deviations is simpler in the case of a non-normal stable limit law (a typical example is the case of power tails with index $\alpha_1 < 2$) than in the case of the normal limit law; compare formulations of Theorems 1.2.1 and 1.2.2. Correspondingly, in this chapter we were able to construct a series of asymptotic expansions for $\mathbf{P}\{S_n > y\}$ having increasing accuracy (i.e., taking into account increasing number of the initial terms of the alternating sum (0.8)) only for the case of power tails with index $\alpha_1 < 2$. For this, we set additional restrictions on the asymptotics of the left-hand tail of $F(x)$. One of possible natural constraints, the fulfilment of which essentially simplifies the construction of a series of asymptotic expansions for $\mathbf{P}\{S_n > y\}$ having increasing accuracy, is condition (0.5′) (the left-hand side analog of (0.5)). Note that for the special case when only the two initial terms of the alternating sum (0.8) are taken into account, we obtain the expansion coinciding with that of Theorem 1.2.1. However, the expansion of Theorem 1.2.1 is valid under milder constraints on the asymptotics of the left-hand tail of $F(x)$.

The main result of this chapter is Theorem 3.1.1, which is formulated and proved in Section 3.1. The proof of this theorem is based on Proposition 3.1.1 which is also formulated in Section 3.1. The proof of Proposition 3.1.1, due to its cumbersomeness, is deferred to Section 3.3. In addition, a series of auxiliary lemmas is also proved in Section 3.1. Section 3.2 contains an illustrative Example 3.2.1 to Theorem 3.1.1. Now, let us proceed with

3.1 Recursive Construction of Asymptotic Expansions of P{S_n > y} in the Case of a Non-Normal Stable Law.

First of all, note that during the derivation of the asymptotic expansions for $\mathbf{P}\{S_n > y\}$ of this section (having increasing accuracy) coefficients possessing certain properties of the standard moments will emerge in some refining terms. For convenience, we introduce these coefficients in advance by means of the following result:

Lemma 3.1.1 Let the tails of the distribution function $F(x)$ of the random variable η satisfy Conditions (0.5) - (0.5′), with its own set of $\{\alpha_i, c_{\alpha_i}, d_{\alpha_i}, r\}$, and all α_i's and r be non-integers.

Then for any non-negative integer $k \leq [r]$ the following integrals converge being understood as improper Stieltjes integrals:

$$\int_{-\infty}^{0} x^k \cdot d\Big(F_\eta(x) - \sum_{l:\alpha_l<k} d_{\alpha_l} \cdot |x|^{-\alpha_l}\Big); \tag{3.1.1}$$

$$\int_{0}^{\infty} x^k \cdot d\Big(F_\eta(x) - 1 + \sum_{l:\alpha_l<k} c_{\alpha_l} \cdot |x|^{-\alpha_l}\Big). \tag{3.1.1′}$$

Proof of Lemma 3.1.1 consists of integration of both integrals by parts with a subsequent application of relationships (0.5) - (0.5′) to the expressions within the braces in (3.1.1) and (3.1.1′), both in the neighborhood of zero, and as $x \to \infty$. □

Let us denote the sum of the integrals (3.1.1) and (3.1.1′) by $b_k(\eta)$. Hereinafter, we refer to $b_k(\eta)$ as k^{th} pseudomoment in Cramér's sense of r.v. η (H. Cramér (1963) implicitly used the coefficients $b_k(X_1)$ for the construction of the asymptotic expansion of the distribution function of r.v. $S_n/n^{1/\alpha_1}$ in the range of normal deviations, under fulfilment of certain conditions analogous to (0.5) and (0.5′), cf. pp. 18 - 19 therein.) It is easily seen that if r.v. η possesses the finite k^{th} moment (i.e. if $k \leq [\alpha_1]$) then it coincides with $b_k(\eta)$.

Obviously, an arbitrary random variable possesses the finite zero-order pseudomoment in Cramér's sense equal to 1.

Let us emphasize that a number of various concepts of pseudomoments is known (cf., e.g., Chapter 2 of the present monograph, Ibragimov and Linnik (1971) Chapter XIV, Section 5, Cristoph (1982) p. 70, and in particular Christoph and Wolf (1992) Chapter 2). The above introduced pseudomoments in Cramér's sense differ from others. In this respect, note that the various pseudomoments considered in Chapter 2 of Christoph and Wolf (1992) imply the closeness (up to a certain accuracy) of the tails of $F(x)$ and of the tails of the limiting stable distribution function (see also formula (3.2.4) of the next section). In turn, conditions of such type imply that d.f. F satisfies (0.5) - (0.5′) with the specific values of the parameters α_i, c_α, and d_α that are stipulated by the asymptotic behavior of the limiting stable distribution on the tails, given by (3.2.4′). See also Example 3.2.1 of the next section for more details. On the other hand, we pose quite different in character and much weaker conditions, which constitute the closeness of the tails of d.f. $F(x)$ to certain linear combinations of power functions that might not correspond to *any* distribution.

Now, let us introduce the following functions for positive real x and s; $s \le r$:

$$\Delta^{+}_{<s}(\eta,x) = \mathbf{P}\{\eta > x\} - \sum_{i:\alpha_i<s} c_{\alpha_i} x^{-\alpha_i};$$

$$\Delta^{-}_{<s}(\eta,x) = \mathbf{P}\{\eta < -x\} - \sum_{i:\alpha_i<s} d_{\alpha_i} x^{-\alpha_i}.$$

The functions $\Delta^{+}_{\le s}(\eta,x)$ and $\Delta^{-}_{\le s}(\eta,x)$ are defined by the same formulas with the change of strict inequalities to non-strict ones in the sums over i in the above formulas.

Using this notation, we obtain that for any r.v. η with the tails satisfying Conditions (0.5) and (0.5′) with a certain set $\{\alpha_i, c_\alpha, d_\alpha, r\}$ such that all α_i's and r are not integers, the following representation is valid:

$$(3.1.2) \qquad b_k(\eta) = \int_{-\infty}^{0} x^k \cdot d\Delta^-_{<k}(\eta,|x|) + \int_{0}^{+\infty} x^k \cdot d\Delta^+_{<k}(\eta,x).$$

In addition, let us introduce the following condition that is related to a special case $r \leq 1$:

$$(3.1.3) \qquad \int_{x}^{\infty} |d\Delta^-_{\leq r}(\eta,v)| + \int_{x}^{\infty} |d\Delta^+_{\leq r}(\eta,v)| = o(x^{-r})$$

as $x \to \infty$. Note that Condition (3.1.3) is analogous to Condition (B′) introduced by Inzhevitov (1986).

The following lemma demonstrates that under fulfilment of (0.5) and (0.5′), and for any integer $1 \leq k \leq [r]$ (if $r > 1$) the coefficient under $(it)^k/k!$ in the expansion of the characteristic function $f_\eta(t)$ of r.v. η in the neighborhood of zero coincides with $b_k(\eta)$, i.e., pseudomoments in Cramér's sense possess the important property of the standard moments.

Lemma 3.1.2. Let the tails of the distribution function $F_\eta(x)$ of r.v. η satisfy Conditions (0.5) - (0.5′) with a certain set of $\{\alpha_i, c_{\alpha_i}, d_{\alpha_i}, r\}$, and all α_i's and r be non-integers. For the case $r < 1$ we also assume the fulfilment of Condition (3.1.3). Then

$$(3.1.4) \qquad f_\eta(t) = \sum_{k=0}^{[r]} b_k(\eta)\cdot(it)^k/k! + \sum_{l=1}^{\ell} R_t(\alpha_l,c_{\alpha_l},d_{\alpha_l})\cdot|t|^{\alpha_l} + o(|t|^r)$$

as $|t| \to 0$, where

$$(3.1.5) \qquad R_t(\theta,a,b) := \Gamma(1-\theta)\cdot\left((a+b)\cdot\cos(\frac{\pi}{2}\cdot t) - i\cdot(a-b)\cdot sgnt\cdot\sin(\frac{\pi}{2}\cdot t)\right).$$

Proof of Lemma 3.1.2 is very simple and therefore is omitted. It is analogous to the proof of the validity of relationship (3.3) of Cramér (1963) if we take into account the remark on p. 107 of Inzhevitov (1986). Note that in the case $r < 1$, the first sum on the right-hand side of (3.1.4) consists of only one term equal to 1. Note that in (3.1.5), we use the standard continuation of the gamma-function onto $\mathbf{C} \setminus \{0; -1; -2; \ldots\}$ by means of the formula $\Gamma(z) = \Gamma(z+1)/z$. □

Let us point out that in Lemmas 3.1.1 and 3.1.2 we do not require that $\alpha_l < 2$ (in contrast to the works by Cramér (1963) and Inzhevitov (1986)).

Now, we proceed with the investigation of the distributions of the sums of r.v.'s $\{X_n, n \geq 1\}$. Set $k_* := [r/\alpha_1] + 1$, and $r_* := k_* \cdot \alpha_1$. Obviously, $r < r_* \leq r + \alpha_1$. In the present section, we will give the asymptotic expansion for $\mathbf{P}\{S_n > y\}$, which takes into account the k_* initial terms of the alternating sum (0.8) (see Theorem 3.1.1 below). The pseudomoments in Cramér's sense of r.v.'s $S_{n-1}, S_{n-2}, \ldots$ will emerge as factors of some refining terms of this expansion. Let us point out that in order to make this expansion more valuable it would be desirable to express these factors, $b_k(S_{n-\ell})$, in terms of the constants known à priori, namely, pseudomoments in Cramér's sense of r.v.'s X_i.

Let us assume the fulfilment of the following condition:

(3.1.6) all non-trivial linear combinations of $\{\alpha_1, \ldots, \alpha_\ell\}$ with non-negative integer coefficients which are less than or equal to r_*, are not integers.

The next lemma demonstrates that under the above condition, the formulas for pseudomoments take on the simplest form.

Lemma 3.1.3. Assume that the sum S_n possesses the finite pseudomoments in Cramér's sense up to the pth. Then $b_1(S_n), \ldots, b_p(S_n)$ are expressed in terms of $b_1(X_1), \ldots, b_p(X_1)$ by the same formulas as the standard moments, i.e. for any integer $1 \leq k \leq p$,

$$b_k(S_n) = k! \cdot \sum_{\substack{(m_1,\ldots,m_k): \\ m_1+2m_2+\ldots+km_k=k}} n^{m_1+\ldots+m_k} \cdot \prod_{\ell=1}^{k} \frac{1}{m_\ell!} \cdot \left(\frac{\kappa_\ell(X_1)}{\ell!}\right)^{m_\ell}, \tag{3.1.7}$$

where for any integer $1 \leq k \leq [r]$ the coefficient $\kappa_\ell(X_1)$ (which is an analog of the semi-invariant) is expressed in terms of the moments by the following formula:

$$\kappa_\ell(X_1) = k! \cdot \sum_{\substack{(m_1,\ldots,m_k): \\ m_1+2m_2+\ldots+km_k=k}} (-1)^{m_1+\ldots+m_k-1} \cdot (m_1+\ldots+m_k-1)! \cdot \prod_{\ell=1}^{k} \frac{1}{m_\ell!} \cdot \left(\frac{b_\ell(X_1)}{\ell!}\right)^{m_\ell}. \tag{3.1.8}$$

Note that the summation in formulas (3.1.7) - (3.1.8) is carried out over all non-negative integer solutions $(m_1,..., m_k)$ to the equation $m_1 + 2m + ... + km_k = k$.

Proof of Lemma 3.1.3. First, let us apply Lemma 3.1.2. Then relationship (3.1.4) yields an analogous relationship concerning the asymptotic behavior of $\log f(t)$ as $|t| \to 0$ with the accuracy up to $o(|t|^r)$. Note that the fulfilment of Condition (3.1.6) enables one to neglect the summands containing non-integer powers of (it) at all, when operating with summands containing non-negative integer powers of (it), since integer and non-integer powers are not mixing up. Thereupon, by analogy with Lemma 2 of Cramér (1963), we use the following relationship:

$$f_{S_n}(t) = \exp\{n \cdot \log f_{X_1}(t)\}.$$

As the result, we obtain the asymptotic representation for $f_{S_n}(t)$ as $|t| \to 0$, which is analogous to (3.1.4), and the existence of finite pseudomoments in Cramér's sense, up to the pth one, for the sum S_n yields that the coefficient under $(it)^m/m!$ in this relationship coincides with $b_m(S_n)$ for any integer $1 \leq m \leq p$. This makes clear the fact that $\{b_m(S_n)\}$ are expressed in terms of $\{b_m(X_1)\}$ by the same formulas as the standard moments. Note that the formulas concerning relationships between moments can be found in Petrov (1975a) (cf. Chapter I, formula (3.3) and Chapter VI, Lemma 1 and formula (1.6) therein); recall that our formulas (3.1.7) - (3.1.8) are their analogs for pseudomoments in Cramér's sense. □

The following result is easily obtained by applying Lemma 3.1.3:

Lemma 3.1.4. Let Conditions (0.5), (0.5′), and (3.1.6) be fulfilled with $\alpha_1 \in (0,1) \cup (1,2)$ and non-integer $r > 1$; $\mathbf{E}X_1 = 0$ if $\alpha_1 > 1$. Assume also that for any integer $n \geq 1$, the sum S_n possesses the finite pseudomoments in Cramér's sense up to the pth one ($1 \leq p \leq [r]$). Then

i) if $\alpha_1 \in (0,1)$ then for any integer $1 \leq k \leq p$, the pseudomoment in Cramér's sense $b_k(S_n)$ is a polynomial on n, whose power does not exceed k;

ii) if $\alpha_1 \in (1,2)$ then for any integer $2 \leq k \leq p$, the pseudomoment in Cramér's sense $b_k(S_n)$ is a polynomial on n, whose power does not exceed $[k/2]$;

the same assertion is also valid in the case $\alpha_1 \in (0,1)$ if $b_1(X_1) = 0$.

Proof of Lemma 3.1.4 narrows down to the determination of the maximum of the sum $m_1 + m_2 + ... + m_k$ under the restriction $m_1 + 2m_2 + ... + km_k = k$ with the subsequent application of Lemma 3.1.3. □

Now, in order to formulate the main result of this chapter, we introduce some additional notation. Let us denote by $\{\beta_j\}$ the set of positive numbers that do not exceed r_* and can be represented in the form

$$\beta_j = \sum_i m_{ji}\cdot\alpha_i + n_j,$$

where m_{ji} and n_j are non-negative integers and

$$\sum_i m_{ji} > 0.$$

Set $\theta := 1/(2[r/\alpha_1]+3)$. For real $x > 0$ we also define the following functions:

$$\Delta_+(x) := \mathbf{P}\{X_i > x\} - \sum_{i=1}^{t} c_{\alpha_i}\cdot x^{-\alpha_i};$$

$$\Delta_-(x) := \mathbf{P}\{X_i < -x\} - \sum_{i=1}^{t} d_{\alpha_i}\cdot x^{-\alpha_i},$$

(Clearly, $\Delta_+(x) = \Delta^+_{\le r}(X_1;x)$, and $\Delta_-(x) = \Delta^-_{\le r}(X_1;x)$.)

Theorem 3.1.1. Let Conditions (0.5), (0.5′), and (3.1.6) be satisfied with $\alpha_1 \in (0,1) \cup (1,2)$; $\mathbf{E}X_i = 0$ if $\alpha_1 > 1$. Then there exist sets of polynomials $\{c_{\beta_j}(n)\}$ and $\{d_{\beta_j}(n)\}$, whose powers do not exceed $[\beta_j/\alpha_1]$ such that

$$\mathbf{P}\{S_n > y\} = \sum_j c_{\beta_j}(n)\cdot y^{-\beta_j} + r_1^+(n,y) + r_2^+(n,y);$$

$$\mathbf{P}\{S_n < -y\} = \sum_j d_{\beta_j}(n)\cdot y^{-\beta_j} + r_1^-(n,y) + r_2^-(n,y),$$

where the remainders $r^+_1(n,y)$ and $r^-_1(n,y)$ are such that there exists a function $K(\cdot)$ from $\mathbf{R}_+^1$ into $\mathbf{R}_+^1$ such that for any positive $\varepsilon > 0$, for any integer $n \ge 1$, and for $y \ge K(\varepsilon)n^{1/\alpha_1}$,

$$|r_1^{\pm}(n,y)| \le \varepsilon\cdot(n\cdot y^{-\alpha_1})^{k_*}$$

(recall that $k_* = [r/\alpha_1] + 1$), and the second remainders are such that there exist constants K_1 and K_2, such that for any integer $n \ge 1$, and for $y \ge K_1\cdot n^{1/\alpha_1}$,

$$|r_2^+(n,y)| \le K_2 \cdot n \cdot \sup_{x \ge \theta y} |\Delta_+(x)|;$$

$$|r_2^-(n,y)| \le K_2 \cdot n \cdot \sup_{x \le -\theta y} |\Delta_-(x)|.$$

Note that some of the polynomials $c_\beta(n)$ and $d_\beta(n)$ can be equal to zero identically.

Remark 3.1.1. For each of the remainders $r_1^\pm(n,y)$ and $r_2^\pm(n,y)$, there exists its own respective range of values of parameters y and n, where it absorbs the other remainder (with the same upper index). For the sake of simplicity, we compare the remainders $r_1^+(n,y)$ and $r_2^+(n,y)$ for the special case $n \to \infty$. It is easily seen that

$$r_2^+(n,y) = o(n \cdot y^{-r})$$

by Condition (0.5). On the other hand,

$$r_1^+(n,y) = o\left((n \cdot y^{-\alpha_1})^{[r/\alpha_1]+1}\right).$$

Thus, the remainder

$$r_2^+(n,y) = o\left((n \cdot y^{-\alpha_1})^{[r/\alpha_1]+1}\right)$$

in the range of deviations

$$y = o\left(n^{[r/\alpha_1]/(r_* - r)}\right).$$

In contrast to this, in the range of deviations

$$y / n^{[r/\alpha_1]/(r_* - r)} \to \infty$$

we get that $r^+_1(n,y) = o(n \cdot y^{-r})$.

Proof of Theorem 3.1.1 is the most cumbersome part of this monograph. Note that in its proof, we use the already proved Lemmas 1.1.1 - 1.1.3 and 3.1.4, and also a series of other auxiliary results, namely, Lemmas 3.1.5 - 3.1.10, 3.1.8′, and Proposition 3.1.1. The proof of Proposition 3.1.1 is routine and relies on the construction of the asymptotic expansions for $\mathbf{P}\{S_n > y\}$ and $\mathbf{P}\{S_n < -y\}$ having the increasing accuracy. This proof is deferred to Section 3.3 due to its awkwardness and also to the fact that it is not used in other parts of this chapter.

The initial stage of the proof of Theorem 3.1.1 follows along the same lines as that of Theorem 1.1.1. We apply relationship (1.1.2) with $u = \theta\cdot y$; the first summand on the right-hand side of (1.1.2) is proved to be negligible, and we can drop all the terms of the alternating sum over m on the right-hand side of (1.1.2) for which $m > k_* \wedge n$. Obviously, the assertions of the theorem in the case $n = 1$ coincide with Conditions (0.5) and (0.5′). Hence, we can hereinafter assume that $n \geq 2$.

First we assume that $n > k_*$ and then apply relationship (1.1.2) with $u = \theta\cdot y$. Then

$$\mathbf{P}\{S_n > y\} - \sum_{m=1}^{k_*}(-1)^{m+1}\cdot\binom{n}{m}\cdot\mathbf{P}\{S_n > y, X_{n-m+1} > \theta\cdot y, \ldots, X_n > \theta\cdot y\}$$
$$= \mathbf{P}\{S_n > y, \max(X_1, \ldots, X_n) \leq \theta\cdot y\} + \sum_{m=k_*+1}^{n}(-1)^{m+1}\cdot\binom{n}{m} \tag{3.1.9}$$
$$\cdot\,\mathbf{P}\{S_n > y, X_{n-m+1} > \theta\cdot y, \ldots, X_n > \theta\cdot y\}.$$

We estimate the first and the second terms on the right-hand side of (3.1.9) by the use of Lemmas 1.1.1 and 1.1.3 with $u = \theta\cdot y$ and $\ell = k_* + 1$, and by the use of Lemma 1.1.2 with $u = \theta\cdot y$ and $k = k_* + 1$, respectively. As the result we obtain that there exist positive constants K_1 and K_2, such that for any integer $n > k_*$ and for $y \geq K_1\cdot n^{1/\alpha_1}$,

$$\left|\mathbf{P}\{S_n > y\} - \sum_{m=1}^{k_*}(-1)^{m+1}\cdot\binom{n}{m}\cdot\mathbf{P}\{S_n > y, X_{n-m+1} > \theta\cdot y, \ldots, X_n > \theta\cdot y\}\right|$$
$$\leq K_2\cdot(n\cdot y^{-\alpha_1})^{k_*+1} + \binom{n}{k_*+1}\cdot\mathbf{P}\{S_n > y, X_{n-k_*} > \theta\cdot y, \ldots, X_n > \theta\cdot y\}. \tag{3.1.10}$$

In order to estimate the rightmost term of this inequality, we use Condition (0.5) along with the fact that $X_1, \ldots, X_n$ are independent and identically distributed r.v.'s. We get that there exists a positive constant K_3, such that for $y \geq K_1\cdot n^{1/\alpha_1}$,

$$\binom{n}{k_*+1}\cdot\mathbf{P}\{S_n > y, X_{n-k_*} > \theta\cdot y, \ldots, X_n > \theta\cdot y\} \leq K_3\cdot(n\cdot y^{-\alpha_1})^{k_*+1}.$$

The latter inequality along with (3.1.10) implies that

$$\mathbf{P}\{S_n > y\} = \sum_{m=1}^{k_*} (-1)^{m+1} \cdot \binom{n}{m} \cdot \mathbf{P}\{S_n > y, X_{n-m+1} > \theta \cdot y, \ldots, X_n > \theta \cdot y\} \Big| \tag{3.1.11}$$
$$+ \ \delta_1(n,y).$$

where the remainder $\delta_1(n,y)$ is such that for any integer $n > k_*$ and for $y \geq K_1 \cdot n^{1/\alpha_1}$,

$$|\delta_1(n,y)| \leq (K_2 + K_3) \cdot (ny^{-\alpha_1})^{k_*+1}.$$

Now, let $n \leq k_*$. Once again, let us apply equality (1.1.2) with $u = \theta \cdot y$. We get that

$$\mathbf{P}\{S_n > y\} = \sum_{m=1}^{n} (-1)^{m+1} \cdot \binom{n}{m} \cdot \mathbf{P}\{S_n > y, X_{n-m+1} > \theta \cdot y, \ldots, X_n > \theta \cdot y\} \tag{3.1.12}$$
$$+ \ \mathbf{P}\{S_n > y, \max(X_1, \ldots, X_n) \leq \theta \cdot y\}.$$

Note the following: the fact that $\max(X_1,\ldots,X_n) \leq \theta \cdot y$ implies that $S_n \leq n \cdot \theta \cdot y < y/2$. Therefore, in this case $\mathbf{P}\{S_n > y, \max(X_1,\ldots,X_n) \leq \theta \cdot y\} = 0$. The latter equality along with (3.1.12) implies that for any integer $1 \leq n \leq k_*$ and for any real $y > 0$,

$$\mathbf{P}\{S_n > y\} = \sum_{m=1}^{n} (-1)^{m+1} \cdot \binom{n}{m} \cdot \mathbf{P}\{S_n > y, X_{n-m+1} > \theta \cdot y, \ldots, X_n > \theta \cdot y\}. \tag{3.1.13}$$

Combining (3.1.11) and (3.1.13) we get that

$$\mathbf{P}\{S_n > y\} = \sum_{m=1}^{k_*} (-1)^{m+1} \cdot \binom{n}{m} \cdot \mathbf{P}\{S_n > y, X_{n-m+1} > \theta \cdot y, \ldots, X_n > \theta \cdot y\} \tag{3.1.14}$$
$$+ \ \delta_2(n,y),$$

where the remainder $\delta_2(\cdot,\cdot)$ is such that there exist positive constants K_1 and K_2 such that for any integer $n \geq 1$ and for any $y \geq K_1 \cdot n^{1/\alpha_1}$,

$$|\delta_2(n,y)| \leq K_2(n \cdot y^{-\alpha_1})^{k_*+1}.$$

Let us denote the probabilities contained in the sum on the right-hand side of (3.1.14) by P_m ($= P_m(n)$):

$$P_m := \mathbf{P}\{S_n > y, X_{n-m+1} > \theta \cdot y, \ldots, X_n > \theta \cdot y\},$$

where $1 \leq m \leq k_* \wedge n$.

Let us proceed with the study of P_m. It is easily seen that for any integer $1 \leq m \leq k_* \wedge (n-1)$,

$$(3.1.15)\qquad \begin{aligned} P_m &= \mathbf{P}\{S_{n-m} > (1-m\theta)\cdot y, X_{n-m+1} > \theta\cdot y, \ldots, X_n > \theta\cdot y\} \\ &+ \int_{-\infty}^{(1-m\theta)y} \mathbf{P}\{X_{n-m+1} > \theta y, \ldots, X_n > \theta y, X_{n-m+1} + \ldots + X_n > y - z\}\cdot dF_{S_{n-m}}(z), \end{aligned}$$

where $F_{S_{n-m}}(\cdot)$ is the distribution function of the sum S_{n-m}. To simplify the above representation for P_m, we introduce the family of functions Π_m (= $\Pi_m(y,z)$), where m is a non-negative integer, $m \le k_*$, $y > 0$, $z \in \mathbf{R}^1$, as follows:

$\Pi_0 :\equiv 1$, and

$\Pi_m(y,z) := \mathbf{P}\{X_1 > \theta y, \ldots, X_m > \theta y;\ S_m \le y - z\}$.

It is obvious that for any integer $1 \le m \le k_*$, for any real $y > 0$, and for $z \ge (1 - m\theta)y$, $\Pi_m(y,z) = 0$. An application of the definition of the functions Π_m and the fact that r.v.'s $X_1, \ldots, X_n$ are independent and identically distributed, enables one to represent the integrand in the rightmost term on the right-hand side of (3.1.15) as the following difference:

$$\begin{aligned} &\mathbf{P}\{X_{n-m+1} > \theta y, \ldots, X_n > \theta y, X_{n-m+1} + \ldots + X_n > y - z\} \\ &\quad = \mathbf{P}\{X_{n-m+1} > \theta y, \ldots, X_n > \theta y\} - \mathbf{P}\{X_{n-m+1} > \theta y, \ldots, X_n > \theta y, \\ &X_{n-m+1} + \ldots + X_n \le y - z\} \\ &\quad = \mathbf{P}\{X_1 > \theta y\}^m - \Pi_m(y,z). \end{aligned}$$

These equalities along with (3.1.15) imply that for any integer $1 \le m \le k_* \wedge (n-1)$,

$$(3.1.16)\qquad P_m = \mathbf{P}\{X_1 > \theta y\}^m - \int_{-\infty}^{(1-m\theta)y} \Pi_m(y,z)\cdot dF_{S_{n-m}}(z).$$

In addition, it is obvious that for any real $y > 0$,

$$(3.1.16')\qquad P_n = \mathbf{P}\{X_1 > \theta y\}^n - \Pi_n(y,0).$$

First, we demonstrate that both first summands on the right-hand sides of (3.1.16) - (3.1.16′) are properly approximated by the corresponding powers of the expression

$$\sum_{i=1}^{l} c_{\alpha_i}\cdot(\theta y)^{-\alpha_i}.$$

Making simple transformations we obtain the result that for any integer $m \ge 2$,

$$\mathbf{P}\{X_1 > \theta y\}^m - \left(\sum_{i=1}^{\ell} c_{\alpha_i}\cdot(\theta y)^{-\alpha_i}\right)^m = \left(\mathbf{P}\{X_1 > \theta y\} - \sum_{i=1}^{\ell} c_{\alpha_i}\cdot(\theta y)^{-\alpha_i}\right)$$

$$\cdot\left(\mathbf{P}\{X_1 > \theta y\}^{m-1} + \mathbf{P}\{X_1 > \theta y\}^{m-2}\cdot\sum_{i=1}^{\ell} c_{\alpha_i}\cdot(\theta y)^{-\alpha_i} + \ldots + \left(\sum_{i=1}^{\ell} c_{\alpha_i}\cdot(\theta y)^{-\alpha_i}\right)^{m-1}\right).$$

This equality along with Condition (0.5) implies that for any integer $m \geq 1$,

$$|\mathbf{P}\{X_1 > \theta y\}^m - \left(\sum_{i=1}^{\ell} c_{\alpha_i}\cdot(\theta y)^{-\alpha_i}\right)^m| \leq \left(\mathbf{P}\{X_1 > \theta y\} + |\sum_{i=1}^{\ell} c_{\alpha_i}\cdot(\theta y)^{-\alpha_i}|\right)^{m-1}$$

$$\cdot \sup_{x \geq \theta y} |\Delta_+(x)|.$$

The latter estimate along with the fact that the fulfilment of Condition (0.5) implies the fulfilment of Condition (0.10), yields that there exists a positive constant K such that for any integer $m \geq 1$ and for any real $y > 0$,

$$|\mathbf{P}\{X_1 > \theta y\}^m - \left(\sum_{i=1}^{\ell} c_{\alpha_i}\cdot(\theta y)^{-\alpha_i}\right)^m| \leq (K\cdot y^{-\alpha_1})^{m-1}\cdot\sup_{x \geq \theta y}|\Delta_+(x)|. \tag{3.1.17}$$

In order to study the second terms on the right-hand sides of (3.1.16) - (3.1.16′), we first investigate the properties of the functions Π_m, where $m \geq 1$ is an arbitrary integer, $m \leq k_*$. To this end, let us derive the recursive formula for these functions. Using the definition of Π_{m+1} we get that

$$\begin{aligned}\Pi_{m+1}(y,z) = \int_{\theta y}^{(1-m\theta)y-z} dF_{X_{m+1}}(v)\cdot\Bigg\{&\int_{\theta y}^{(1-(m-1)\theta)y-z-v} dF_{X_m}(z_1)\\ &\cdot\int_{\theta y}^{(1-(m-2)\theta)y-z-v-z_1} dF_{X_{m-1}}(z_2)\cdot\ldots\cdot\int_{\theta y}^{y-z-v-z_1-\ldots-z_{m-1}} dF_{X_1}(z_m)\Bigg\};\end{aligned} \tag{3.1.18}$$

for the case $m = 0$ the expression within the braces is assumed to be 1. Then for any non-negative integer $m < k_*$, we obtain that the expression within the braces is equal to $\Pi_m(y,z+v)$. Therefore, we derive the following recursive version of the definition of the family $\{\Pi_m\}$ for $0 \leq m \leq k_*$:

(3.1.18′)

$$\Pi_0 := 1,$$

$$\Pi_{m+1}(y,z) = \int_{\theta y}^{(1-m\theta)y-z} \Pi_m(y,z+v)\cdot dF(v)$$

for $z < (1-(m+1)\theta)y$, and $\Pi_{m+1} \equiv 0$ for $z \geq (1-(m+1)\theta)y$.

Now, let us sketch the plan of investigation of the integrals on the right-hand side of (3.1.16):

$$(3.1.19) \qquad \int_{\theta y}^{(1-m\theta)y} \Pi_m(y,z)\cdot dF_{S_{n-m}}(z),$$

where $1 \leq m \leq k_* \wedge (n-1)$. Recall that the event whose probability of realization is equal to $\Pi_m(y,z)$ is related to the values of r.v.'s $X_1,\ldots, X_m$ that are greater than θy. Hence, it is not very difficult to compute Π_m up to the known asymptotics of the right-hand tail of $F(x)$. Let us emphasize that the computation up to such accuracy is possible due to the fact that we deal with the distribution of a finite number (not greater than k_*) of the random variables taking large values. On the other hand, in order to study the properties of the distribution function $F_{S_{n-m}}$ in the differential in (3.1.19), a different approach should be chosen, on the reason that $m \leq k_*$, and n may not turn to infinity. Namely, the already known results on the asymptotics of the tails of the distribution function of the sum S_{n-m} should be used. Applying these results to studying the integrals (3.1.19), we can get the asymptotics of the tails of $F_{S_{n-m}}$ up to a better accuracy, etc. This approach is realized in Proposition 3.1.1, which is proved in Section 3.3.

Now, we proceed with pursuing the first part of our program, namely, the derivation of the asymptotics of the functions Π_m. First, we recursively introduce the family of functions $\mathbf{H}_m$ ($= \mathbf{H}_m(y,z)$), where $0 \leq m \leq k_*$, $y > 0$, and $z \in \mathbb{R}_+^1$:

$$\mathbf{H}_0 \;:=\; 0;$$

$$\mathbf{H}_{m+1} := \int_{\theta y}^{(1-m\theta)y-z} \mathbf{H}_m(y,z+v)\cdot d\Big(-\sum_{i=1}^{\ell} c_{\alpha_i}\cdot(\theta y)^{-\alpha_i}\Big) \text{ if } z < (1-(m+1)\theta)y;$$

$$\mathbf{H}_{m+1} := \quad 0 \qquad\qquad \text{if } z \geq (1-(m+1)\theta)y.$$

Note that just in the same way as the expression

$$\sum_{i=1}^{\ell} c_{\alpha_i}\cdot(\theta y)^{-\alpha_i}$$

is an approximation for the right-hand tail of the distribution function $F(x)$ depending on a finite number of parameters and smooth (in contrast to the function F), the functions $\mathbf{H}_m$ are approximations of Π_m depending on a finite number of parameters and smooth. It is easy to demonstrate that for any integer $1 \leq m \leq k_*$, the function $\mathbf{H}_m$ is continuous in the second argument over the whole real axis, because $\mathbf{H}_m(y,z) \to 0$ as $z \uparrow (1\text{-}m\theta)y$, but its partial derivative in the second argument is discontinuous at the point $z = (1\text{-}m\theta)y$. Note that the definition of the functions $\mathbf{H}_m(y,z)$ is analogous to the recursive version of the definition of the functions $\Pi_m(y,z)$. Now, set $t := z/y$ and introduce the function $\mathbf{h}_m(y,t)$ as follows:

$$\mathbf{h}_m(y,t) := H_m(y,z) \qquad \text{if } t < 1 - m\cdot\theta;$$

$$\mathbf{h}_m(y,t) := 0 \qquad\qquad \text{if } t \geq 1 - m\cdot\theta.$$

Thus, we get that $\mathbf{h}_0 \equiv 0$; the function $\mathbf{h}_{m+1}(y,t)$ is equal to zero for $t \geq 1$ - $(m+1)\theta$, and the following recursive formula is fulfilled for $t < 1$ - $(m+1)\cdot\theta$:

$$\text{(3.1.20)} \qquad \mathbf{h}_{m+1}(t) = \sum_{i=1}^{\ell} c_{\alpha_i}\cdot(\theta y)^{-\alpha_i}\cdot \int_{\theta}^{1-m\theta-t} \mathbf{h}_m(y,t+\tau)\cdot d(-\tau^{-\alpha_i}).$$

Recall that the function Π_{m+1} has been introduced by means of formula (3.1.18). It is clear that the computation of Π_{m+1} is based on the integration over the set of $(m+1)$ variables, which in turn enables one to derive the recursive representation (3.1.18′). In contrast to that, we already have the recursive representation (3.1.20) for the function $\mathbf{h}_{m+1}(y,t)$. Starting from this relationship, we now derive the formula for $\mathbf{h}_m(y,t)$ that is analogous to (3.1.18) and is based on the integration over the set of m variables. To this end, we introduce a family of functions $g_{i_1,\ldots,i_m}(t)$ with the indices $i_1, \ldots, i_m$ taking all integer values between 1 and ℓ, and $t \in (-\infty, 1 - m\theta)$:

$$g_{i_1,\dots,i_m}(t) := \int\limits_0^{1-(m-1)\theta-t} d\left(-\tau_1^{-\alpha_{i_1}}\right) \cdot \int\limits_0^{1-(m-2)\theta-t-\tau_1} d\left(-\tau_2^{-\alpha_{i_2}}\right)$$

$$\cdot \dots \cdot \int\limits_0^{1-t-\tau_1-\dots-\tau_{m-1}} d\left(-\tau_m^{-\alpha_{i_m}}\right).$$

Recall that $1 \le m \le k_*$. Obviously, the functions $g_{i_1,\dots,i_m}(\cdot)$ are infinitely differentiable on the ray $(-\infty, 1 - m\theta)$.

Using this notation, we obtain from (3.1.20) that for $t \in (-\infty, 1 - m\theta)$,

$$\mathbf{h}_m(y,t) = \sum_{\substack{(i_1,\dots,i_m):\\ 1 \le i_1,\dots,i_m \le \ell}} c_{\alpha_{i_1}} \cdot \dots \cdot c_{\alpha_{i_m}} \cdot y^{-\alpha_{i_1}-\dots-\alpha_{i_m}} \cdot g_{i_1,\dots,i_m}(t), \tag{3.1.20$'$}$$

where the summation is carried out over all sets of non-negative integer numbers such that $1 \le i_1, \dots, i_m \le \ell$.

In particular, this relationship implies the infinite differentiability of the functions $\mathbf{H}_m(y,z)$ in the second argument for $z \in (-\infty, (1 - m\theta)y)$, since for any non-negative integer s,

$$\frac{\partial^s \mathbf{H}_m(y,z)}{\partial z^s} = \frac{1}{y^s} \cdot \frac{\partial^s \mathbf{h}_m(y,t)}{\partial t^s}, \tag{3.1.21}$$

where $t = z/y$. Besides, the following lemma is valid:

Lemma 3.1.5. Let the following inequality be fulfilled for any non-negative integer $m \le k_*$, for any real $x > 0$, and for any non-negative integer p:

$$\left|\frac{\partial^p \mathbf{H}_m(y,z)}{\partial z^p}\right| \le K \cdot y^{-m\alpha_1 - p} \tag{3.1.22}$$

uniformly in $z \in (-\infty, (1 - m\theta)y)$, where K is some positive constant that does not depend on m and on p.

Proof of Lemma 3.1.5. It follows from (3.1.20′) and (3.1.21) that for $z \in (-\infty, (1 - m\theta)y)$,

$$(3.1.23)\qquad \frac{\partial^p \mathbf{H}_m(y,z)}{\partial z^p} = \sum_{\substack{(i_1,\dots,i_m):\\ 1\le i_1,\dots,i_m\le l}} c_{\alpha_{i_1}}\cdot\ldots\cdot c_{\alpha_{i_m}}\cdot y^{-\alpha_{i_1}-\dots-\alpha_{i_m}-p}\cdot g^{(p)}_{i_1,\dots,i_m}(t),$$

where $t = z/y$. The rest of the proof is trivial. □

The next lemma enables one to replace the function $\Pi_m(y,z)$ by its smooth approximation $\mathbf{H}_m(y,z)$ during the evaluation of the integral (3.1.19).

Lemma 3.1.6. For any non-negative integer $m \le k_*$ and for any real $y > 0$,

$$(3.1.24)\qquad |\Pi_m(y,z) - \mathbf{H}_m(y,z)| \le K\cdot y^{-(m-1)\alpha_1}\cdot \sup_{x\ge\theta y}|\Delta_+(x)|$$

uniformly in $z \in \mathbf{R}^1$, where K is some positive constant that can depend on m.

Proof of Lemma 3.1.6 is carried out by induction on m. The induction base follows from the fact that $\mathbf{H}_0 \equiv \Pi_0 \equiv 1$; let us prove the induction step. First, we note that it suffices to establish the validity of estimate (3.1.20) only for $z < (1 - (m + 1)\theta)y$, since for $z \ge (1 - (m + 1)\theta)y$ both functions Π_{m+1} and $\mathbf{H}_{m+1}$ are equal to zero. In addition, it follows from the definition of the functions

$$\Pi_m(y,z) := \mathbf{P}\{X_1 > \theta y, \dots , X_m > \theta y, S_m \le y - z\}$$

that they are monotonically non-increasing functions in the second argument, and $\Pi_m(y,-\infty) = (1 - F(\theta y))^m$ for any integer $1 \le m \le k_*$. Now, let us apply the recursive relationship for the function Π_{m+1} (see (3.1.18′)). We change the function $F(\mathrm{v})$ on the right-hand side of this formula for the following sum:

$$\left(1 - \sum_{i=1}^{l} c_{\alpha_i}\cdot \mathrm{v}^{-\alpha_i}\right) + \left(-\Delta_+(\mathrm{v})\right).$$

The purpose of introducing the function $\Delta_+(\mathrm{v})$ is to compensate the error emerged from the change of $F(\mathrm{v})$ to

$$1 - \sum_{i=1}^{l} c_{\alpha_i}\cdot \mathrm{v}^{-\alpha_i}.$$

Thus, we obtain the following equality:

$$\Pi_{m+1}(y,z) - \mathbf{H}_{m+1}(y,z) := \int_{\theta y}^{(1-m\theta)y-z} \Pi_m(y,z+v)\cdot d\Big(-\Delta_+(v)\Big)$$

(3.1.25)

$$+ \int_{\theta y}^{(1-m\theta)y-z} \Big(\Pi_m(y,z+v) - \mathbf{H}_m(y,z+v)\Big)\cdot d\Big(-\sum_{l=1}^{t} c_{\alpha_l}\cdot(\theta y)^{-\alpha_l}\Big).$$

Integrating the first integral on the right-hand side of (3.1.25) by parts we obtain that this integral is equal to

$$-\Pi_m(y,z+v)\cdot\Delta_+(v)\Big|_{\theta y}^{(1-m\theta)y-z} + \int_{\theta y}^{(1-m\theta)y-z} \Delta_+(v)\cdot d\big(\Pi_m(y,z+v)\big).$$

Recall that the function Π_m is monotonically non-increasing function in the second argument, and that $\Pi_m(y,-\infty) = (1 - F(\theta y))^m$. Therefore, the absolute value of the first integral on the right-hand side of (3.1.25) does not exceed

$$2\cdot\big(1-F(\theta y)\big)^m\cdot\sup_{x\ge\theta y}|\Delta_+(x)| + \sup_{x\ge\theta y}|\Delta_+(x)|\cdot\int_{\theta y}^{(1-m\theta)y-z} d\big(-\Pi_m(y,z+v)\big)$$

(3.1.26)

$$\le 3\cdot\big(1-F(\theta y)\big)^m\cdot\sup_{x\ge\theta y}|\Delta_+(x)|.$$

Now, we derive an analogous estimate for the absolute value of the second integral on the right-hand side of (3.1.25): the absolute value of this integral does not exceed

$$\sum_{l=1}^{t}|c_{\alpha_l}|\cdot\int_{\theta y}^{(1-m\theta)y-z}|\Pi_m(y,z+v) - \mathbf{H}_m(y,z+v)|\cdot d\Big(-v^{-\alpha_l}\Big).$$

In order to get an upper estimate for this sum, we apply the induction hypothesis and also make the use of the fact that the functions $-v^{-\alpha_l}$ are monotonic. We get that the absolute value of the second integral on the right-hand side of (3.1.25) does not exceed

$$\sum_{l=1}^{t}|c_{\alpha_l}|\cdot(\theta y)^{-\alpha_l}\cdot\sup_{z+\theta y\le v\le(1-m\theta)y}|\Pi_m(y,v) - \mathbf{H}_m(y,v)|$$
$$\le K_m\cdot\theta^{\alpha_1}\cdot\sum_{l=1}^{t}|c_{\alpha_l}|\cdot y^{-(m-1)\alpha_1-\alpha_l}\cdot\sup_{x\ge\theta y}|\Delta_+(x)|.$$

Combining this inequality with (3.1.25) and (3.1.26) we obtain the assertion of the induction step. □

Now, we proceed with the other lemma that will be used in the proof of Proposition 3.1.1.

Lemma 3.1.7. There exist positive constants K_1 and K_2 such that for any integer $n \geq 2$, for any integer $1 \leq m \leq k_* \wedge (n\text{-}1)$, and for $y \geq K_1 \cdot n^{1/\alpha_1}$,

$$(3.1.27)\qquad \left|\binom{n}{m}\cdot\left(P_m - \left(\sum_{l=1}^{l} c_{\alpha_l}\cdot(\theta y)^{-\alpha_l}\right)^m + \int_{-\infty}^{(1-m\theta)y} \mathbf{H}_m(y,z)\cdot dF_{S_{n-m}}(z)\right)\right| \leq K_2 \cdot n \cdot \sup_{x\geq\theta y}|\Delta_+(x)|.$$

Proof of Lemma 3.1.7. Recall (see formula (3.1.16)) that for any integer $1 \leq m \leq k_* \wedge (n\text{-}1)$,

$$P_m = \mathbf{P}\{X_1 > \theta y\}^m - \int_{-\infty}^{(1-m\theta)y} \Pi_m(y,z)\cdot dF_{S_{n-m}}(z).$$

The first term is estimated by the use of inequality (3.1.17): for any real $y > 0$,

$$(3.1.28)\qquad \left|P_m - \left(\sum_{l=1}^{l} c_{\alpha_l}\cdot(\theta y)^{-\alpha_l}\right)^m + \int_{-\infty}^{(1-m\theta)y} \Pi_m(y,z)\cdot dF_{S_{n-m}}(z)\right| \leq (K\cdot y^{-\alpha_1})^{m-1}\cdot \sup_{x\geq\theta y}|\Delta_+(x)|,$$

where K is some positive constant. Now, let us estimate the following difference:

$$\int_{-\infty}^{(1-m\theta)y} \Pi_m(y,z)\cdot dF_{S_{n-m}}(z) - \int_{-\infty}^{(1-m\theta)y} \mathbf{H}_m(y,z)\cdot dF_{S_{n-m}}(z).$$

It is easily seen that the absolute value of this expression does not exceed

$$\int_{-\infty}^{(1-m\theta)y} |\Pi_m(y,z) - \mathbf{H}_m(y,z)|\cdot dF_{S_{n-m}}(z) \leq \mathbf{P}\{S_{n-m} \leq (1-m\theta)y\}$$

$$\cdot \sup_{-\infty < z \le (1-m\theta)y} |\Pi_m(y,z) - \mathbf{H}_m(y,z)| \le K\cdot y^{-m\alpha_1}\cdot \sup_{x\ge\theta y} |\Delta_+(x)|$$

by Lemma 3.1.6. Combining these inequalities with (3.1.28) we get that for any integer $n \ge 2$, for any integer $1 \le m \le k_* \wedge (n\text{-}1)$, and for any real $y > 0$,

$$\left| P_m - \left(\sum_{l=1}^{t} c_{\alpha_l}\cdot(\theta y)^{-\alpha_l}\right)^m + \int_{-\infty}^{(1-m\theta)y} H_m(y,z)\cdot dF_{S_{n-m}}(z)\right|$$

$$\le \left(K_3^{m-1} + K_4\right)\cdot y^{-(m-1)\alpha_1}\cdot \sup_{x\ge\theta y} |\Delta_+(x)|.$$

Multiplying both sides of this inequality by $\binom{n}{m}$ turns the left-hand side into the left-hand side of inequality (3.1.27), whereas the right-hand side admits the following estimate:

$$\left(K_3^{m-1} + K_4\right)\cdot\binom{n}{m}\cdot y^{-(m-1)\alpha_1}\cdot \sup_{x\ge\theta y} |\Delta_+(x)|$$

$$\le Const\cdot(n\cdot y^{-\alpha_1})^{m-1}\cdot n\cdot \sup_{x\ge\theta y} |\Delta_+(x)|.$$

It is easily seen that in the range of deviations $y \ge n^{1/\alpha_1}$, the right-hand side of the above inequality does not exceed the right-hand side of (3.1.27). □

Let us emphasize that the problem of studying the integrals (3.1.19) is hence reduced to the problem of studying the following integrals:

$$(3.1.29) \qquad \int_{-\infty}^{(1-m\theta)y} \mathbf{H}_m(y,z)\cdot dF_{S_{n-m}}(z),$$

where $1 \le m \le k_* \wedge (n\text{-}1)$. Note that the smoothness of the functions $\mathbf{H}_m$ enables one to approximate these functions by the corresponding Taylor expansions (taken in some neighborhood of the point $(y,0)$) during the evaluation of the integrals (3.1.29). In this respect, the errors emerged due to such approximations should be estimated, and in order to simplify the formulas we introduce the following notation:

$$\mathbf{H}_m^{(s)}(y,z) \quad := \frac{\partial^s \mathbf{H}_m(y,z)}{\partial z^s};$$

$$\Delta\mathbf{H}_m(y,z,p) := \mathbf{H}_m(y,z) - \sum_{s=0}^{p} \frac{\mathbf{H}_m^{(s)}(y,0)}{s!} \cdot z^s,$$

where $1 \le m \le k_*$; s and p are arbitrary non-negative integer numbers; $y > 0$, and $z < (1-m\theta)y$. Now, let us represent the integral (3.1.29) as the following sum:

$$\mathbf{H}_m(y,0) \cdot \left(1 - \mathbf{P}\{S_{n-m} > (1-m\theta)y\}\right) \tag{3.1.30}$$
$$+ \int_{-\infty}^{(1-m\theta)y} \left(\mathbf{H}_m(y,z) - \mathbf{H}_m(y,0)\right) \cdot dF_{S_{n-m}}(z).$$

Using the just introduced notation, the identity

$$F_{S_{n-m}}(z) \equiv 1 - \mathbf{P}\{S_{n-m} > z\}$$

and the continuity of the function $\mathbf{H}_m(\cdot,\cdot)$ in the second argument we obtain that the integral from (3.1.30) is equal to

$$\int_{-\infty}^{(1-m\theta)y} \Delta\mathbf{H}_m(y,z,0) \cdot dF_{S_{n-m}}(z) \tag{3.1.31}$$
$$= \int_{-\infty}^{0} \Delta\mathbf{H}_m(y,z,0) \cdot d\left(\mathbf{P}\{S_{n-m} < z\}\right) + \int_{0}^{(1-m\theta)y} \Delta\mathbf{H}_m(y,z,0) \cdot d\left(1 - \mathbf{P}\{S_{n-m} > z\}\right),$$

since the continuity of $\Delta\mathbf{H}_m(y,z,0)$ in the second argument along with the fact that $\Delta\mathbf{H}_m(y,0,0) = 0$ enables one to replace $F_{S_{n-m}}(z)$ $(= \mathbf{P}\{S_{n-m} \le z\})$ by $\mathbf{P}\{S_{n-m} < z\}$ in the integral from $-\infty$ to zero on the right-hand side of (3.1.31). Now, (3.1.30), (3.1.31) and Lemma 3.1.7 imply the following result.

Lemma 3.1.8. For any integer $n \ge 2$ and for any integer $1 \le m \le k_* \wedge (n-1)$,

$$\binom{n}{m} \cdot P_m = \binom{n}{m} \cdot \left\{ \left(\sum_{i=1}^{l} c_{\alpha_i} \cdot (\theta y)^{-\alpha_i} \right)^m - \mathbf{H}_m(y,0) \right. \tag{3.1.32}$$

$$+\ \mathbf{H}_m(y,0)\cdot\int_{(1-m\theta)y}^{+\infty} z^0\cdot d\big(-\mathbf{P}\{S_{n-m}>z\}\big)\ -\ \int_{-\infty}^{0}\Delta\mathbf{H}_m(y,z,0)\cdot d\mathbf{P}\{S_{n-m}<z\}$$

$$-\int_0^{(1-m\theta)y}\Delta\mathbf{H}_m(y,z,0)\cdot d\big(-\mathbf{P}\{S_{n-m}>z\}\big)\Big)\Bigg\}\ +\ \Delta_m(n,y),$$

where the remainder $\Delta_m(\cdot,\cdot)$ is such that there exist positive constants K_1 and K_2 such that in the range of deviations $y \geq n^{1/\alpha_1}$,

(3.1.33) $$|\Delta_m(n,y)|\ \leq\ K_2\cdot n\cdot\sup_{x\geq\theta y}|\Delta_+(x)|.$$

Proof of Lemma 3.1.8 is trivial, since relationships (3.1.32) and (3.1.33) easily follow from (3.1.30), (3.1.31), Lemma 3.1.7 (see relationship (3.1.27)), and the fact that the leftmost integral on the right-hand side of (3.1.32) coincides with $\mathbf{P}\{S_{n-m} > (1-m\theta)y\}$. □

The following result provides similar relationships for $\binom{n}{n}\cdot P_n$:

Lemma 3.1.8′. Let $n \leq k_*$. Then

(3.1.32′) $$\binom{n}{n}\cdot P_n\ =\ \Big(\sum_{i=1}^{l} c_{\alpha_i}\cdot(\theta y)^{-\alpha_i}\Big)^n\ -\ \mathbf{H}_n(y,0)\ +\ \Delta_n(n,y),$$

where the remainder $\Delta_n(\cdot,\cdot)$ is such that there exists a positive constant K such that for any real $y \geq 1$,

(3.1.33′) $$|\Delta_n(n,y)|\ \leq\ K\cdot n\cdot\sup_{x\geq\theta y}|\Delta_+(x)|.$$

Proof of Lemma 3.1.8′ is very simple. Combining (3.1.16′), (3.1.17), and (3.1.24) we get that for any real $y > 0$,

$$\Big|P_n\ -\ \Big(\sum_{i=1}^{l} c_{\alpha_i}\cdot(\theta y)^{-\alpha_i}\Big)^n\ +\ \mathbf{H}_n(y,0)\Big|\ \leq\ Const\cdot y^{-(n-1)\alpha_1}\cdot\sup_{x\geq\theta y}|\Delta_+(x)|.$$

It only remains to note that $\binom{n}{n} = 1$. □

Let us note that the comparison of the assertions of Lemmas 3.1.8 and 3.1.8′ demonstrates that in the case $n > k_*$ the representations for some of the quantities $\binom{n}{m} \cdot P_m$, where $m \le k_* \wedge (n-1)$, are rather cumbersome, as they are given in terms of three summands comprised of various integrals. The following lemma (that will be used in the proof of Proposition 3.1.1) enables one to neglect the three latter integrals from the expression within the braces on the right-hand side of (3.1.32) in some cases.

Lemma 3.1.9. There exists a function $\mathbf{L}_0(\cdot)$ from $\mathbf{R}_+^1$ to $\mathbf{R}_+^1$ such that for any integer $n \ge 2$, for any integer $1 \le m \le k_* \wedge (n-1)$, for any real $\varepsilon > 0$, and for $y \ge \mathbf{L}_0(\varepsilon)\, n^{1/\alpha_1}$,

$$\left| \binom{n}{m} \cdot \mathbf{H}_m(y,0) \cdot \int_{(1-m\theta)y}^{+\infty} z^0 \cdot d\left(-\mathbf{P}\{S_{n-m} < z\}\right) \right| \le \epsilon \cdot (n \cdot y^{-\alpha_1})^m, \tag{3.1.34}$$

$$\left| \binom{n}{m} \cdot \int_{-\infty}^{0} \Delta\mathbf{H}_m(y,z,0) \cdot d\mathbf{P}\{S_{n-m} < z\} \right| \le \epsilon \cdot (n \cdot y^{-\alpha_1})^m, \tag{3.1.34′}$$

$$\left| \binom{n}{m} \cdot \int_{0}^{(1-m\theta)y} \Delta\mathbf{H}_m(y,z,0) \cdot d\left(-\mathbf{P}\{S_{n-m} > z\}\right) \right| \le \epsilon \cdot (n \cdot y^{-\alpha_1})^m. \tag{3.1.34′′}$$

Proof of Lemma 3.1.9. Let us establish the validity of estimates (3.1.34), (3.1.34′), and (3.1.34′′) with $\mathbf{L}_0(\varepsilon) = K\varepsilon^{-2/\alpha_1}$, where K is some positive constant.

First, inequality (3.1.34) easily follows from Remark 1.1.1 to Proposition 1.1.1 and Lemma 3.1.5.

Now, we proceed with the derivation of (3.1.34). To this end, we fix an arbitrary positive ε and split the integral from $-\infty$ to zero into the sum of the integrals from $-\infty$ to $-\varepsilon$ (I_1) and from $-\varepsilon$ to zero (I_2). The former one is easily estimated by applying Lemma 3.1.5:

$$\left| \binom{n}{m} \cdot I_1 \right| \le \binom{n}{m} \cdot K_1 \cdot y^{-m\alpha_1} \cdot \mathbf{P}\{S_{n-m} < -\epsilon y\},$$

where K_1 is some positive constant. Using this inequality along with Proposition 1.1.1 we derive that

$$\left|\binom{n}{m}\cdot \mathrm{I}_1\right| \le (n\cdot y^{-\alpha_1})^m \cdot K_2\cdot n\cdot(\epsilon\cdot y)^{-\alpha_1}.$$

Obviously, for $y \ge K_2^{1/\alpha_1}\cdot \varepsilon^{-2/\alpha_1}\cdot n^{1/\alpha_1}$, the inequality

$$K_2\cdot n\cdot(\epsilon\cdot y)^{-\alpha_1} \le \epsilon$$

is valid, and hence

$$\left|\binom{n}{m}\cdot \mathrm{I}_1\right| \le \epsilon\cdot(n\cdot y^{-\alpha_1})^m. \tag{3.1.35}$$

In order to estimate the integral I_2, recall that

$$\Delta\mathbf{H}_m(y,z,0) := \mathbf{H}_m(y,z) - \mathbf{H}_m(y,0).$$

By Lagrange's theorem, this difference is equal to $\mathbf{H}_m^{(1)}(y,v)\cdot z$, where $z \in [-\varepsilon,0)$, and $v \in (z,0)$. Estimating $|\mathbf{H}_m^{(1)}(y,v)|$ by the use of Lemma 3.1.5 we obtain this result:

$$\left|\binom{n}{m}\cdot \mathrm{I}_2\right| \le K_3\cdot(n\cdot y^{-\alpha_1})^m\cdot\frac{1}{y}\cdot\epsilon. \tag{3.1.35′}$$

It only remains to note that (3.1.35) and (3.1.35′) imply (3.1.34′). The derivation of inequality (3.1.34′′) follows along the same lines as that of inequality (3.1.34′) and therefore is omitted. □

Note that in the proof of Proposition 3.1.1, the formulas for $\mathbf{H}_m^{(s)}(y,0)$, as well as for the integrals

$$\int_0^{(1-m\theta)y} \Delta\mathbf{H}_m(y,z,[\gamma])\cdot d(-z^{-\gamma})$$

and

$$\int_{-\infty}^{0} \Delta\mathbf{H}_m(y,z,[\gamma])\cdot d|z|^{-\gamma},$$

$$\int_{-\infty}^{0} \Delta\mathbf{H}_m(y,z,[\gamma])\cdot d|z|^{-\gamma},$$

where the parameter γ takes on positive non-integer values, will be required. Such formulas are easily obtained by applying relationship (3.1.23):

Lemma 3.1.10. For any integer $1 \le m \le k_*$, for any non-negative integer s, and for any positive non-integer γ,

$$\text{(3.1.36)} \qquad \mathbf{H}_m^{(s)}(y,0) = \sum_{\substack{(i_1,\dots,i_m):\\ 1\le i_1,\dots,i_m\le \ell}} c_{\alpha_{i_1}}\cdot\ldots\cdot c_{\alpha_{i_m}}\cdot y^{-\alpha_{i_1}-\dots-\alpha_{i_m}-s}\cdot g_{i_1,\dots,i_m}^{(s)}(0);$$

$$\text{(3.1.36}'\text{)} \qquad \int_0^{(1-m\theta)y} \Delta\mathbf{H}_m(y,z,[\gamma])\cdot d(-z^{-\gamma}) = \sum_{\substack{(i_1,\dots,i_m):\\ 1\le i_1,\dots,i_m\le \ell}} c_{\alpha_{i_1}}\cdot\ldots\cdot c_{\alpha_{i_m}}\cdot y^{-\alpha_{i_1}-\dots-\alpha_{i_m}-\gamma}$$
$$\cdot \int_0^{1-m\theta}\left(g_{i_1,\dots,i_m}(t) - \sum_{s=0}^{[\gamma]}\frac{g_{i_1,\dots,i_m}^{(s)}(0)}{s!}\cdot t^s\right)\cdot d(-t^{-\gamma});$$

$$\text{(3.1.36}''\text{)} \qquad \int_{-\infty}^{0} \Delta\mathbf{H}_m(y,z,[\gamma])\cdot d|z|^{-\gamma} = \sum_{\substack{(i_1,\dots,i_m):\\ 1\le i_1,\dots,i_m\le \ell}} c_{\alpha_{i_1}}\cdot\ldots\cdot c_{\alpha_{i_m}}\cdot y^{-\alpha_{i_1}-\dots-\alpha_{i_m}-\gamma}$$
$$\cdot \int_{-\infty}^{0}\left(g_{i_1,\dots,i_m}(t) - \sum_{s=0}^{[\gamma]}\frac{g_{i_1,\dots,i_m}^{(s)}(0)}{s!}\cdot t^s\right)\cdot d|t|^{-\gamma}.$$

Proof of Lemma 3.1.10 is not at all difficult. It relies on relationship (3.1.23) and the well-known formulas on the change of variables in the integral. □

Now, we proceed with the construction of the asymptotic expansions for $\mathbf{P}\{S_n > y\}$ and $\mathbf{P}\{S_n < -y\}$ having increasing accuracy. Recall that the most precise of them are given in Theorem 3.1.1. It is convenient to present recursive formulas for these expansions in advance. To this end, we now introduce some auxiliary notation.

For arbitrary non-negative integers k and n we set $\mathrm{I}_{\{k\le n\}} := 1$ if $k \le n$ and $\mathrm{I}_{\{k\le n\}} := 0$ otherwise. Now, let us introduce recursively the four families of functions: $\Sigma_\pm^{(k)}(n,y)$ and $\Sigma_\pm^{(k)}(k_1,n,y)$, where k and k_1 are non-negative integers that do not exceed k_* and $k_* + 1$, respectively, and $k < k_1$. The functions of the families $\Sigma_\pm^{(k)}(n,y)$ will be represented by the sums of the products of polynomials on n and negative powers of y, which are greater than or equal to $-k\alpha_1$; these functions will serve as approximations for $\mathbf{P}\{S_n > y\}$ and $\mathbf{P}\{S_n < -y\}$. It will be shown during the construction of these two families that the above sums can be represented in the following form:

$$\Sigma_{+}^{(k)}(n,y) = \sum_{j:\beta_j \le k\alpha_1} c_{\beta_j}(n) \cdot y^{-\beta_j}, \tag{3.1.37}$$

$$\Sigma_{-}^{(k)}(n,y) = \sum_{j:\beta_j \le k\alpha_1} d_{\beta_j}(n) \cdot y^{-\beta_j}, \tag{3.1.37$'$}$$

where β_j can be represented as

$$\beta_j = \sum_j m_{ji} \cdot \alpha_i + n_j,$$

m_{ji} and n_j are non-negative integer numbers;

$$\sum_j m_{ji} > 0,$$

and $c_{\beta_j}(n)$ and $d_{\beta_j}(n)$ are the polynomials on n of the orders not exceeding $[\beta_j/\alpha_1]$. The representations of the functions $\Sigma_{\pm}^{(k)}(n,y)$ by means of (3.1.37) - (3.1.37′) will be also used for the construction of subsequent terms of the two other families. Note that the functions from the third and the fourth families will be also the sums of the products of polynomials on n and negative powers of y, that are greater than or equal to $-k_1\alpha_1$, but less than $-k\alpha_1$.

Now, let us proceed with the recursive construction of these four families. The initial functions for all the four families are set to be zero:

$$\Sigma_{\pm}^{(0)}(n,y) :\equiv \Sigma_{\pm}^{(0)}(k_1,n,y) :\equiv 0.$$

The recursive relationships for these families are as follows:

$$\begin{aligned} \Sigma_{+}^{(k+1)}(n,y) :=\; & \Sigma_{+}^{(k)}(n,y) + \Sigma_{+}^{(k)}(k+1,n,y) + \mathrm{I}_{\{k+1 \le n\}} \cdot (-1)^{k+2} \cdot c_{\alpha_1}^{k+1} \\ & \cdot \left(\theta^{-(k+1)\alpha_1} - g_{1,\dots,1}(0)\right) \cdot \binom{n}{k+1} \cdot y^{-(k+1)\cdot\alpha_1} + \sum_{m=1}^{k\wedge(n-1)} (-1)^{m+1} \cdot \binom{n}{m} \\ & \cdot \left(\sum_{s=0}^{[(k-m)\alpha_1]} \frac{1}{s!} \cdot \sum_{j:(k-m)\alpha_1 < \beta_j \le (k+1-m)\alpha_1} (1-m\theta)^{s-\beta_j} \cdot \frac{\beta_j}{\beta_j - s} \cdot c_{\beta_j}(n-m) \right. \end{aligned} \tag{3.1.38}$$

$$\cdot \sum_{\substack{(i_1,\dots,i_m):\\ 1\le i_1,\dots,i_m\le l;\\ \beta_j+\alpha_{i_1}+\dots+\alpha_{i_m}\le(k+1)\alpha_1}} c_{\alpha_{i_1}}\cdot\ldots\cdot c_{\alpha_{i_m}}\cdot g^{(s)}_{i_1,\dots,i_m}(0)\cdot y^{-\beta_j-\alpha_{i_1}-\dots-\alpha_{i_m}}$$

$$+\sum_{s=[(k-m)\alpha_1]+1}^{[(k+1-m)\alpha_1]}\frac{1}{s!}\cdot\sum_{j:\ s<\beta_j\le(k+1-m)\alpha_1}(1-m\theta)^{s-\beta_j}\frac{\beta_j}{\beta_j-s}c_{\beta_j}(n-m)$$

$$\cdot \sum_{\substack{(i_1,\dots,i_m):\\ 1\le i_1,\dots,i_m\le l:\\ \beta_j+\alpha_{i_1}+\dots+\alpha_{i_m}\le(k+1)\alpha_1}} c_{\alpha_{i_1}}\cdot\ldots\cdot c_{\alpha_{i_m}} g^{(s)}_{i_1,\dots,i_m}(0) y^{-\beta_j-\alpha_{i_1}-\dots-\alpha_{i_m}}$$

$$-\sum_{s=[(k-m)\alpha_1]+1}^{[(k+1-m)\alpha_1]}\frac{b_s(S_{n-m})}{s!}\sum_{\substack{(i_1,\dots,i_m):\\ 1\le i_1,\dots,i_m\le l;\\ s+\alpha_{i_1}+\dots+\alpha_{i_m}\le(k+1)\alpha_1}} c_{\alpha_{i_1}}\cdot\ldots\cdot c_{\alpha_{i_m}}\cdot g^{(s)}_{i_1,\dots,i_m}(0)\cdot y^{-s-\alpha_{i_1}-\dots-\alpha_{i_m}}$$

$$-\sum_{j:\ (k-m)\alpha_1<\beta_j\le(k+1-m)\alpha_1} c_{\beta_j}(n-m)\cdot\sum_{\substack{(i_1,\dots,i_m):\\ 1\le i_1,\dots,i_m\le l;\\ \beta_j+\alpha_{i_1}+\dots+\alpha_{i_m}\le(k+1)\alpha_1}} c_{\alpha_{i_1}}\cdot\ldots\cdot c_{\alpha_{i_m}}$$

$$\cdot y^{-\beta_j-\alpha_{i_1}-\dots-\alpha_{i_m}}\cdot\int_0^{1-m\theta}\left(g_{i_1,\dots,i_m}(t)-\sum_{s=0}^{[\beta_j]}\frac{g^{(s)}_{i_1,\dots,i_m}(0)}{s!}\cdot t^s\right)\cdot d\left(-t^{-\beta_j}\right)$$

$$-\sum_{j:\ (k-m)\alpha_1<\beta_j\le(k+1-m)\alpha_1} d_{\beta_j}(n-m)\cdot\sum_{\substack{(i_1,\dots,i_m):\\ 1\le i_1,\dots,i_m\le l;\\ \beta_j+\alpha_{i_1}+\dots+\alpha_{i_m}\le(k+1)\alpha_1}} c_{\alpha_{i_1}}\cdot\ldots\cdot c_{\alpha_{i_m}}$$

$$\cdot y^{-\beta_j-\alpha_{i_1}-\dots-\alpha_{i_m}}\cdot\int_{-\infty}^{0}\left(g_{i_1,\dots,i_m}(t)-\sum_{s=0}^{[\beta_j]}\frac{g^{(s)}_{i_1,\dots,i_m}(0)}{s!}\cdot t^s\right)\cdot d\,|t|^{-\beta_j}\Bigg);$$

$$
\begin{aligned}
&\Sigma_+^{(k+1)}(k_1,n,y) := \Sigma_+^{(k)}(k_1,n,y) - \mathrm{I}_{\{k_1 \le k+1\}} \cdot \Sigma_+^{(k)}(k+1,n,y) + \mathrm{I}_{\{k+1 \le n\}} \\
&\cdot (-1)^{k+2} \cdot \sum_{\substack{(i_1,\dots,i_{k+1}):\\ 1 \le i_1,\dots,i_{k+1} \le \ell;\\ i_1 \wedge \dots \wedge i_{k+1} > 1\\ \alpha_{i_1}+\dots+\alpha_{i_{k+1}} \le k_1\alpha_1}} c_{\alpha_i} \cdot \ldots \cdot c_{\alpha_{i_{k+1}}} \cdot \left(\theta^{-\alpha_{i_1}-\dots-\alpha_{i_{k+1}}} - g_{i_1,\dots,i_{k+1}}(0)\right) \\
&\cdot \binom{n}{k+1} \cdot y^{-\alpha_{i_1}-\dots-\alpha_{i_{k+1}}} + \sum_{m=1}^{k\wedge(n-1)} (-1)^{m+1} \cdot \binom{n}{m} \cdot \Bigg(\sum_{s=0}^{[(k-m)\alpha_1]} \frac{1}{s!} \\
&\cdot \sum_{j:\,(k-m)\alpha_1 < \beta_j \le (k+1-m)\alpha_1} (1-m\theta)^{s-\beta_j} \cdot \frac{\beta_j}{\beta_j - s} \cdot c_{\beta_j}(n-m) \cdot \sum_{\substack{(i_1,\dots,i_m):\\ 1 \le i_1,\dots,i_{k+1} \le \ell;\\ (k+1)\alpha_1 < \beta_j+\alpha_{i_1}+\dots+\alpha_{i_m} \le k_1\alpha_1}} \\
&c_{\alpha_{i_1}} \cdot \ldots \cdot c_{\alpha_{i_m}} \cdot g_{i_1,\dots,i_m}^{(s)}(0) \cdot y^{-\beta_j-\alpha_{i_1}-\dots-\alpha_{i_m}} - \sum_{s=[(k-m)\alpha_1]+1}^{[(k+1-m)\alpha_1]} \frac{b_s(S_{n-m})}{s!} \\
&\cdot \sum_{\substack{(i_1,\dots,i_m):\\ 1 \le i_1,\dots,i_m \le \ell;\\ (k+1)\alpha_1 < s+\alpha_{i_1}+\dots+\alpha_{i_m} \le k_1\alpha_1}} c_{\alpha_i} \cdot \ldots \cdot c_{\alpha_{i_m}} \cdot g_{i_1,\dots,i_m}^{(s)}(0) \cdot y^{-s-\alpha_{i_1}-\dots-\alpha_{i_m}} \\
&- \sum_{j:\,(k-m)\alpha_1 < \beta_j \le (k+1-m)\alpha_1} c_{\beta_j}(n-m) \cdot \sum_{\substack{(i_1,\dots,i_m):\\ 1 \le i_1,\dots,i_m \le \ell;\\ (k+1)\alpha_1 < \beta_j+\alpha_{i_1}+\dots+\alpha_{i_m} \le k_1\alpha_1}} c_{\alpha_{i_1}} \cdot \ldots \cdot c_{\alpha_{i_m}} \cdot y^{-\beta_j-\alpha_{i_1}-\dots-\alpha_{i_m}} \\
&\cdot \int_0^{1-m\theta} \left(g_{i_1,\dots,i_m}(t) - \sum_{s=0}^{[\beta_j]} \frac{g_{i_1,\dots,i_m}^{(s)}(0)}{s!} \cdot t^s \right) \cdot d\left(-t^{\beta_j}\right) - \sum_{j:\,(k-m)\alpha_1 < \beta_j \le (k+1-m)\alpha_1} \\
&d_{\beta_j}(n-m) \cdot \sum_{\substack{(i_1,\dots,i_m):\\ 1 \le i_1,\dots,i_m \le \ell;\\ (k+1)\alpha_1 < \beta_j+\alpha_{i_1}+\dots+\alpha_{i_m} \le k_1\alpha_1}} c_{\alpha_{i_1}} \cdot \ldots \cdot c_{\alpha_{i_m}} \cdot y^{-\beta_j-\alpha_{i_1}-\dots-\alpha_{i_m}}
\end{aligned}
\tag{3.1.38'}
$$

$$\cdot \int_{-\infty}^{0} \left(g_{i_1,\ldots,j_m}(t) - \sum_{s=0}^{[\beta_j]} \frac{g_{i_1,\ldots,j_m}^{(s)}(0)}{s!} \cdot t^s \right) \cdot d\,|t|^{-\beta_j} \Bigg).$$

Note that in the case $k = 0$, the sums over m in formulas (3.1.38) and (3.1.38′) are assumed to be zero, and in the case $[(k+1-m)\alpha_1] = [(k-m)\alpha_1]$, the sums over s from $[(k-m)\alpha_1] + 1$ to $[(k+1-m)\alpha_1]$ are also assumed to be zero.

Let us also note that in the recursive formulas (3.1.38) and (3.1.38′) for $\Sigma_+^{(k+1)}(n,y)$ and $\Sigma_+^{(k+1)}(k_1,n,y)$, we use representations (3.1.37) and (3.1.37′) for $\Sigma_\pm^{(1)}(n,y),\ldots, \Sigma_\pm^{(k)}(k_1,n,y)$. Let us emphasize that due to the symmetry between Conditions (0.5) and (0.5′), the recursive formulas analogous to (3.1.38) and (3.1.38′) are valid for $\Sigma_-^{(k+1)}(n,y)$ and $\Sigma_-^{(k+1)}(k_1,n,y)$ as well. Moreover, the symmetry between Conditions (0.5) and (0.5′) makes it possible to use these formulas not even writing them down explicitly.

It follows from (3.1.38) - (3.1.38′) and their left-hand analogs that the functions $\Sigma_\pm^{(1)}(n,y)$ can be represented by formulas (3.1.37) and (3.1.37′) with $c_{\beta_j}(n)$ and $d_{\beta_j}(n)$ being the polynomials on n whose powers do not exceed $[\beta_j/\alpha_1]$. This fact can be established by induction. It is trivial when considering the case of refining terms on the right-hand side of (3.1.38) having the factors of the form $\binom{n}{m} \cdot y^{-\beta_j}$, where $\beta_j > (k-m)\cdot\alpha_1$. On the other hand, for the refining terms having the factors $b_s(S_{n-m})y^s$ Lemma 3.1.4 should be used. We then obtain that $b_s(S_{n-m})$ is a polynomial on n whose power does not exceed s if $\alpha_1 \in (0,1)$ or $[s/2]$ if $\alpha_1 \in (1,2)$. In addition, Condition (3.1.6) implies that all the powers of y that emerge in expressions (3.1.38) and (3.1.38′) are not integer numbers.

Now, for any integer $n \geq 1$, for any real $y > 0$, and for any non-negative integer $k < k_*$, we set

$$\Delta_+(n,y,k) := \mathbf{P}\{S_n > y\} - \Sigma_+^{(k)}(n,y);$$

$$\Delta_-(n,y,k) := \mathbf{P}\{S_n < -y\} - \Sigma_-^{(k)}(n,y).$$

Let us also introduce the following three supplementary families of integrals which will be used in the proof of Proposition 3.1.1:

$$\mathrm{I}_1(m,n-m,k-m) := \int_{-\infty}^{0} \Delta\mathbf{H}_m(y,z,[(k-m)\alpha_1])\cdot d\Delta_-(n-m,-z,k-m);$$

$$\mathrm{I}_2(m,n-m,k-m) := \int_{0}^{(1-m\theta)y} \Delta\mathbf{H}_m(y,z,[(k-m)\alpha_1])\cdot d\left(-\Delta_+(n-m,z,k-m)\right);$$

$$\mathrm{I}(s,m,n-m,k-m) := \int_{(1-m\theta)y}^{\infty} z^s\cdot d\left(-\Delta_+(n-m,z,k-m)\right).$$

Proposition 3.1.1. Assume that all the conditions of Theorem 3.1.1 are fulfilled. Then the following assertions are valid:

i) for any non-negative integer $k \le k_* - 1 = [r/\alpha_1]$ there exists a function $\mathbf{L}_k(\cdot)$ from $\mathbf{R}_+^1$ to $\mathbf{R}_+^1$ such that for any integer $n \ge 2$, for any real $\varepsilon > 0$, and for $y \ge \mathbf{L}_k(\varepsilon)\cdot n^{1/\alpha_1}$,

(3.1.39) $$|\Delta_+(n,y,k)| \le \epsilon\cdot\left(n\cdot y^{-\alpha_1}\right)^k;$$

(3.1.39′) $$|\Delta_-(n,y,k)| \le \epsilon\cdot\left(n\cdot y^{-\alpha_1}\right)^k;$$

ii) for any integer $n \ge 2$ and for $1 \le k \le k_*$, the following assertions are fulfilled:

ii1) $$\Delta_+(n,y,k) = \sum_{m=1}^{k\wedge(n-1)} (-1)^{m+1}\cdot\binom{n}{m}\cdot\left\{\mathrm{I}_1(m,n-m,k-m) + \mathrm{I}_2(m,n-m,k-m)\right\}$$

(3.1.40) $$+ \sum_{m=1}^{k\wedge(n-1)} (-1)^{m+1}\cdot\binom{n}{m}\cdot\sum_{s=0}^{[(k-m)\alpha_1]}\frac{\mathbf{H}_m^{(s)}(y,0)}{s!}\cdot\mathrm{I}(s,m,n-m,k-m)$$

$$+ \Sigma_+^{(k)}(k_*+1,n,y) + \Delta_{1,k}(n,y) + \Delta_{2,k}(n,y) + \Delta_{3,k}(n,y),$$

where

$$\Delta_{1,k}(n,y) := \sum_{m=k+1}^{n} (-1)^{m+1}\cdot\binom{n}{m}\cdot P_m \qquad \text{if } k < n;$$

$$\Delta_{1,k}(n,y) := 0 \qquad \text{otherwise},$$

and the remainders $\Delta_{2,k}(n,y)$ and $\Delta_{3,k}(n,y)$ are such that there exist positive constants K_1, K_2, and K_3 (depending on k) such that for any integer $n \ge 2$, for any integer $1 \le k \le k_*$, and for

$y \geq K_1 n^{1/\alpha_1}$,

$$|\Delta_{2,k}(n,y)| \leq K_2 \cdot \left(n \cdot y^{-\alpha_1}\right)^{k_*+1};$$

$$|\Delta_{3,k}(n,y)| \leq K_3 \cdot n \cdot \sup_{x \geq \theta y} |\Delta_+(x)|;$$

ii2) there exists a function $\mathbf{R}_{1,k}(\cdot)$ from $\mathbf{R}_+^1$ to $\mathbf{R}_+^1$ such that for any integer $n \geq 2$, for any integer $1 \leq m \leq k \wedge (n-1)$, for any real $\varepsilon > 0$, and for $y \geq \mathbf{R}_{1,k} \cdot n^{1/\alpha_1}$,

$$(3.1.41) \qquad \left|\binom{n}{m} \cdot \sum_{s=0}^{[(k-m)\alpha_1]} \frac{\mathbf{H}_m^{(s)}(y,0)}{s!} \cdot \mathrm{I}(s,m,n-m,k-m)\right| \leq \epsilon \cdot \left(n \cdot y^{-\alpha_1}\right)^k;$$

ii3) there exists a function $\mathbf{R}_{2,k}(\cdot)$ from $\mathbf{R}_+^1$ to $\mathbf{R}_+^1$ such that for any integer $n \geq 2$, for any integer $1 \leq m \leq k \wedge (n-1)$, for any real $\varepsilon > 0$, and for $y \geq \mathbf{R}_{2,k} n^{1/\alpha_1}$,

$$(3.1.42) \qquad \left|\binom{n}{m} \cdot \mathrm{I}_1(m,n-m,k-m)\right| \leq \epsilon \cdot \left(n \cdot y^{-\alpha_1}\right)^k;$$

$$(3.1.42') \qquad \left|\binom{n}{m} \cdot \mathrm{I}_2(m,n-m,k-m)\right| \leq \epsilon \cdot \left(n \cdot y^{-\alpha_1}\right)^k;$$

iii) for any integer $1 \leq k \leq k_*$, the left-hand analogs of all of the relationships from points (ii1) - (ii3) are valid.

Conclusion of Proof of Theorem 3.1.1. Note that due to the symmetry between Conditions (0.5) and (0.5′), it suffices to establish only the required relationship for $\mathbf{P}\{S_n > y\}$. Let us apply relationship (3.1.40) from point (ii) of Proposition 3.1.1 with $k := k_*$. Estimating the remainder $\Delta_{1,k_*}(n,y)$ by the use of Lemma 1.1.2 and Condition (0.5) we obtain the result that for any integer $n \geq 2$ and for any real $y \geq 1$,

$$(3.1.43) \qquad |\Delta_{1,k_*}(n,y)| \leq K \cdot \left(n \cdot y^{-\alpha_1}\right)^{k_*+1}.$$

Consequently, keeping in mind the recursive relationship (3.1.38′), we get that there exists a function $\mathbf{W}(\cdot)$ from $\mathbf{R}_+^1$ to $\mathbf{R}_+^1$ such that for any integer $n \geq 2$, for any real $\varepsilon > 0$ and for $y \geq \mathbf{W}(\varepsilon) n^{1/\alpha_1}$,

$$(3.1.44) \qquad \left|\Sigma_{+}^{(k_*)}(k_*+1,n,y)\right| \le \epsilon\cdot\left(n\cdot y^{-\alpha_1}\right)^{k_*}.$$

The first and the second sums on the right-hand side of (3.1.40) are estimated by the use of inequalities (3.1.42) - (3.1.42′) and (3.1.41) with $k = k_*$, respectively. Combining the estimates obtained for these sums with inequalities (3.1.43) - (3.1.44) we derive the assertion of Theorem 3.1.1 with

$$\sum_j c_{\beta_j}(n)\cdot y^{-\beta_j} := \Sigma_{+}^{(k_*)}(n,y);$$

$$r_2^{+}(n,y) := \Delta_{3,k_*}(n,y);$$

$$r_1^{+}(n,y) := \sum_{m=1}^{k_*\wedge(n-1)} (-1)^{m+1}\cdot\binom{n}{m}\cdot\left\{I_1(m,n-m,k_*-m) + I_2(m,n-m,k_*-m)\right\}$$

$$+ \sum_{m=1}^{k_*\wedge(n-1)} (-1)^{m+1}\cdot\binom{n}{m}\cdot\sum_{s=0}^{[(k_*-m)\alpha_1]} \frac{\mathbf{H}_m^{(s)}(y,0)}{s!}\cdot I(s,m,n-m,k_*-m)$$

$$+\Sigma_{+}^{(k_*)}(k_*+1,n,y) + \Delta_{1,k_*}(n,y) + \Delta_{2,k_*}(n,y). \quad \square$$

3.2 Asymptotic Expansions of the Probabilities of Large Deviations of S_n and Non-uniform Estimates of Remainders in Limit Theorems on Weak Convergence to Non-normal Stable Laws.

In the previous section, we have obtained the asymptotic expansions for $\mathbf{P}\{S_n > y\}$ by employing the direct probabilistic method. On the other hand, such expansions can also be derived from the limit theorems on normal deviations with non-uniform estimate of the remainder. By these theorems we mean the results analogous to Propositions 1.2.1 - 1.2.3 of section 1.2.1 and Theorems 2.1.1 and 2.2.1 of Chapter 2, but more precise ones (recall that Corollary 1.2.1 only demonstrates that the remainders presented in these propositions have the same order as $\mathbf{P}\{S_n > y\}$, in the range of large deviations of S_n, whereas Theorems 2.1.1 and 2.2.1 provide asymptotics of the probabilities of large deviations only up to equivalence.)

Asymptotic expansions in the integral theorems on non-normal deviations with the exact uniform estimate of the remainder (in the sense of dependence on n) are easily derived for the distributions satisfying Conditions (0.5) and (0.5′) (cf., e.g., Cramér (1963) Theorems 1 - 2, and Vinogracov (1985b) Theorem 1). In order to get the results of such kind, the method of characteristic functions and Lemma 3.1.2 should be used. However, this method turns out to be less effective in the case when non-uniform estimates of the remainder in the integral theorems should be constructed in such a way that the remainder being considered in the range of large deviations would be negligible (compare to certain terms of the asymptotic expansion). In order to obtain non-uniform estimates of such kind, the fulfilment of additional technical constraints (the existence of the density etc.) should be required. At the same time, the method of characteristic functions turns out to be powerful for the derivation of analogous local limit theorems in the case of absolutely continuous distributions whose densities satisfy local analogs of our integral conditions (0.5) - (0.5′) (cf., e.g., Inzhevitov (1986) for details).

For simplicity, let us assume that $\alpha \in (0,1)$. Then it should be mentioned that Bentkus and Bloznelis (1989) obtained the exact non-uniform estimates for the rate of convergence of the normalized sums of i.i.d.r.v.'s to a non-normal stable law by the use of the method of convolutions, under fulfilment of the following conditions:

$$\int_{-\infty}^{+\infty} |x|^{1+\alpha} \cdot \left| d\big(F(x) - G_\alpha(x)\big) \right| < \infty; \tag{3.2.1}$$

$$\mu_1 := \int_{-\infty}^{+\infty} x \cdot d\big(F(x) - G_\alpha(x)\big) = 0, \tag{3.2.2}$$

where $G_\alpha(\cdot)$ is the distribution function of the non-normal stable law with index α, which is the weak limit for the random sequence $F_{S_n/n^{1/\alpha}}$ as $n \to \infty$. Here, the integrals contained in (3.2.1) and (3.2.2) are in fact pseudomoments that differ from those in Cramér's sense, and were mentioned below Lemma 3.1.1. They have applied the just mentioned exact non-uniform estimate of the remainder for the derivation of the following result on the probabilities of large deviations of S_n:

$$(3.2.3)\qquad \mathbf{P}\{S_n > y\} = \left(1 - G_\alpha(y/n^{1/\alpha})\right)\cdot\left(1 + O(y^{-1})\right)$$

as $n \to \infty$, $y/n^{1/\alpha} \to \infty$ (see relationship (1.3) therein). Now, we proceed with the consideration of a particular example for which we will compare the results that can be obtained by the use of the just quoted results by Bentkus and Bloznelis (1989) and by the use of our Theorem 3.1.1.

Example 3.2.1. Let Conditions (3.2.1) - (3.2.2) be fulfilled. For the sake of simplicity we also assume that $X_1 \geq 0$, and $\alpha \in (1/2,2/3) \cup (2/3,1)$. Obviously, Conditions (3.2.1) - (3.2.2) imply that

$$(3.2.4)\qquad F(x) - G_\alpha(x) = o(x^{-\alpha-1})$$

as $x \to \infty$, and the results of Chapter II by Ibragimov and Linnik (1971) imply that the random sequence $S_n/n^{1/\alpha}$ converges to the non-normal stable law with index α which in turn yields the fulfilment of (0.1) with the parameters α and c_α determined by the asymptotic behavior of the right-hand tail of the stable distribution $G_\alpha(\cdot)$. In addition,

$$(3.2.4')\qquad 1 - G_\alpha(x) = \sum_{k=1}^{\infty}(-1)^{k+1}\cdot\frac{c_\alpha^k\cdot\Gamma(1-\alpha)^k}{k!\cdot\Gamma(1-k\alpha)}\cdot x^{-k\alpha}$$

as $x \to \infty$ (cf., e.g., Theorem 2.4.1 of Ibragimov and Linnik (1971)). Combining (3.2.4) and (3.2.4′) we obtain the result that

$$1 - F(x) = c_\alpha\cdot x^{-\alpha} - \frac{c_\alpha^2\cdot\Gamma(1-\alpha)^2}{2\cdot\Gamma(1-2\alpha)}\cdot x^{-2\alpha} + o(x^{-\alpha-1})$$

as $x \to \infty$; the terms of the alternating series on the right-hand side of (3.2.4′) beginning with the third one are absorbed into the remainder, $o(x^{-\alpha-1})$, on the reason that $\alpha > 1/2$ implies that $3\alpha > \alpha + 1$. Set

$$(3.2.5)\qquad c_{2\alpha} := -\frac{c_\alpha^2\cdot\Gamma(1-\alpha)^2}{2\cdot\Gamma(1-2\alpha)}.$$

Using this notation, we obtain the following representation for the right-hand tail of the distribution function $F(\cdot)$ as $x \to \infty$:

$$(3.2.6)\qquad 1 - F(x) = c_\alpha\cdot x^{-\alpha} + c_{2\alpha}\cdot x^{-2\alpha} + o(x^{-\alpha-1}).$$

Then (3.2.6) along with the fact that r.v.'s $\{X_n, n \geq 1\}$ are not negative imply that for

$\alpha \in (1/2,2/3) \cup (2/3,1)$, all the conditions of our Theorem 3.1.1 are fulfilled with $\ell := 2$, $\alpha_1 := \alpha, \alpha_2 := 2\alpha, r := \alpha + 1, k := [(\alpha+1)/\alpha] + 1 = 3$ (the case $\alpha = 2/3$ should be excluded since otherwise Condition (3.1.6) is not fulfilled: in that case $3\alpha = k. \cdot \alpha = 2$ is integer).

Applying Theorem 3.1.1 we obtain that

$$\mathbf{P}\{S_n > y\} = n \cdot \left(c_\alpha \cdot y^{-\alpha} + c_{2\alpha} \cdot y^{-2y}\right) - \binom{n}{2} \cdot \frac{\Gamma(1-\alpha)^2}{\Gamma(1-2\alpha)} \cdot c_\alpha^2 \cdot y^{-2\alpha}$$

$$+ \binom{n}{3} \cdot \frac{\Gamma(1-\alpha)^3}{\Gamma(1-3\alpha)} \cdot c_\alpha^3 \cdot y^{-3\alpha} + n \cdot b_1(S_{n-1}) \cdot y^{-\alpha-1} + o\left((ny^{-\alpha})^3\right) + o(ny^{-\alpha-1})$$

as $n \to \infty$, $y/n^{1/\alpha} \to \infty$. Now, in order to simplify this expression we use relationship (3.2.5) for $c_{2\alpha}$ and the following equality: $b_1(S_{n-1}) = (n-1) \cdot b_1(X_1)$. Then

$$\mathbf{P}\{S_n > y\} = n \cdot c_\alpha \cdot y^{-\alpha} - \frac{n^2 \cdot \Gamma(1-\alpha)^2}{2! \cdot \Gamma(1-2\alpha)} \cdot c_\alpha^2 \cdot y^{-2\alpha} + \frac{n^3 \cdot \Gamma(1-\alpha)^3}{3! \cdot \Gamma(1-3\alpha)}$$

(3.2.7)

$$\cdot\, c_\alpha^3 \cdot y^{-3\alpha} + n \cdot (n-1) \cdot b_1(X_1) \cdot y^{-\alpha-1} + o\left((ny^{-\alpha})^3\right) + o(ny^{-\alpha-1})$$

as $n \to \infty$, $y/n^{1/\alpha} \to \infty$. It is easily seen that (3.2.4) implies that the sum of the three initial terms on the right-hand side of (3.2.7) is equal to

(3.2.8) $$1 - \mathbf{G}_\alpha(y/n^{1/\alpha}) + o\left((ny^{-\alpha})^3\right)$$

as $n \to \infty$, $y/n^{1/\alpha} \to \infty$. Now, let us demonstrate that

(3.2.9) $$b_1(X_1) = 0.$$

In fact, it is easily seen that

$$\int_{-\infty}^{+\infty} x \cdot d\left(F(x) + c_\alpha \cdot x^{-\alpha}\right) = 0,$$

since by Lemma 3.1.2 this integral coincides with the coefficient under (it) in the expansion of the characteristic function

$$\int_0^\infty e^{itx} \cdot dG_\alpha(x)$$

at some neighborhood of zero. Combining this fact with (3.2.2) we obtain that

$$\mu_1 + \int_{-\infty}^{+\infty} x \cdot d\left(G_\alpha(x) + c_\alpha \cdot x^{-\alpha}\right) = \int_{-\infty}^{+\infty} x \cdot d\left(F(x) + c_\alpha \cdot x^{-\alpha}\right) = 0,$$

which in turn yields (3.2.9).

Combining (3.2.7) - (3.2.9) we get that

$$\text{(3.2.10)} \qquad \mathbf{P}\{S_n > y\} = 1 - G_\alpha(y/n^{1/\alpha}) + o\left((ny^{-\alpha})^3\right) + o(ny^{-\alpha-1})$$

as $n \to \infty$, $y/n^{1/\alpha} \to \infty$. The comparison of the representation for $\mathbf{P}\{S_n > y\}$ that can be derived from Theorem 3.1.1 (cf. (3.2.10)) with the representation for $\mathbf{P}\{S_n > y\}$ that is obtained by applying the method of convolutions (cf. (3.2.3)) demonstrates that in the range of deviations $y = o(n^{2/(2\alpha-1)})$, estimate (3.2.3) (which is due to Bentkus and Bloznelis (1989)) is more accurate, whereas in the range of deviations $y \geq Const \cdot n^{2/(2\alpha-1)}$ estimate (3.2.10) (obtained by the use of the direct probabilistic method) is more accurate. However, let us emphasize that our method and the results are applicable to a wider class of distributions (compare to the results by Bentkus and Bloznelis (1989)). In the cases when both our results and the results by Bentkus and Bloznelis (1989) are applicable, our results provide more accurate estimates for greater values of large deviations.

3.3 Proof of Proposition 3.1.1.

First, let us note that due to the symmetry between Conditions (0.5) - (0.5′), the validity of the assertions of point (ii) for some integer $1 \leq k \leq k_*$ implies the validity of their left-hand analogs for the same k, namely, the assertions of point (iii).

Now, let us demonstrate that the fulfilment of the assertions of point (ii) (and hence, of point (iii)) for some integer $1 \leq k \leq k_* - 1$ yields the validity of inequalities (3.1.39) - (3.1.39′) of point (i). Obviously, for $k = 0$ these inequalities follow from Remark 1.1.1 to Proposition 1.1.1.

We prove this by induction; let us assume that for any integer $1 \leq p \leq k - 1$ ($\leq k_* - 1$), the assertions of all three points of Proposition 3.1.1 are valid. Obviously, due to the symmetry between Conditions (0.5) - (0.5′), it suffices to establish the validity of inequality (3.1.39). For this, we first use relationship (3.1.40) for the considered k.

Applying inequalities (3.1.42) - (3.1.42′) for the estimation of the absolute value of the first sum on the right-hand side of (3.1.40), we obtain that for $y \geq \mathbf{R}_{2,k}(\varepsilon) n^{1/\alpha_1}$, its absolute value does not exceed

$$2k\varepsilon(ny^{-\alpha_1})^k.$$

Similarly, we obtain that inequality (3.1.41) implies that for $y \geq \mathbf{R}_{1,k}(\varepsilon) n^{1/\alpha_1}$, the absolute value of the second sum on the right-hand side of (3.1.40) does not exceed

$$k\varepsilon(ny^{-\alpha_1})^k.$$

The corresponding estimate for $|\Sigma_+^{(k)}(k_* + 1, n, y)|$ is obtained following along the same lines, by applying the recursive relationship (3.1.38′).

Now, we derive similar estimates for the absolute values of the remainders $\Delta_{1,k}(n,y)$, $\Delta_{2,k}(n,y)$, and $\Delta_{3,k}(n,y)$. Obviously, the inequality

$$|\Delta_{1,k}(n,y)| \leq K \cdot (n \cdot y^{-\alpha_1})^{k+1}$$

is valid, since for $k < n$ this estimate coincides with inequality (1.1.6), and for $k \geq n$ this remainder is equal to zero by definition. Hence, for $y \geq \varepsilon^{-1/\alpha_1} \cdot n^{1/\alpha_1}$,

(3.3.1) $$|\Delta_{1,k}(n,y)| \leq K_1 \cdot \epsilon \cdot (n \cdot y^{-\alpha_1})^k.$$

The analogous upper bounds for $|\Delta_{2,k}(n,y)|$ and $|\Delta_{3,k}(n,y)|$ are easily derived from the estimates for these absolute values contained in (ii2). In particular, for $y \geq \varepsilon^{-1/\alpha_1} \cdot n^{1/\alpha_1}$,

(3.3.1′) $$|\Delta_{2,k}(n,y)| \leq K_1 \cdot \epsilon \cdot (n \cdot y^{-\alpha_1})^k.$$

In order to estimate $|\Delta_{3,k}(n,y)|$, we should also employ the fact that the expansion for $1 - F(x)$ given by (0.5) implies the existence of a function $X(\cdot)$ from $\mathbf{R}_+^1$ to $\mathbf{R}_+^1$ such that for any real $\varepsilon > 0$ and for $x \geq X(\varepsilon)$,

$$|\Delta_+(x)| \leq \varepsilon \cdot x^{-r}.$$

Therefore, we get that for $y \geq (K_1 \cdot n^{1/\alpha_1}) \vee (X(\varepsilon)/\theta)$,

$$|\Delta_{3,k}(n,y)| \leq K_2 \cdot \varepsilon \cdot n \cdot (\theta y)^{-r} = K_3 \cdot \varepsilon \cdot n \cdot y^{-r}.$$

Let us denote $\mathbf{R}(\varepsilon) := K_1 \vee (X(\varepsilon)/\theta)$. Thereupon, we use the inequality $\alpha_1 \cdot k \leq r$ (recall that $k \leq k_* - 1 \leq r/\alpha_1$) to derive that for any integer $n \geq 2$, for any real $\varepsilon > 0$, and for $y \geq \mathbf{R}(\varepsilon) n^{1/\alpha_1}$,

(3.3.1′′) $$|\Delta_{3,k}(n,y)| \leq K_3 \cdot \varepsilon \cdot n \cdot y^{-r} \leq K_4 \cdot \varepsilon \cdot (n \cdot y^{-\alpha_1})^k.$$

Combining estimates (3.3.1) - (3.3.1′′) for the remainders with the estimates for the first, the second, and the third sums on the right-hand side of (3.1.40) we obtain the validity of inequality (3.1.39).

The rest of the proof of Proposition 3.1.1 is split into two stages. The first one consists of determination of the validity of all the relationships of point (ii) for any integer $1 \le k \le k_* - 1$ (the relationships of point (iii) for the same k are then obtained automatically, and the validity of the inequalities of point (i) for any integer $1 \le k \le k_* - 1$ follows from the results related to point (ii), by the already established part of Proposition 3.1.1). Note that this stage of the proof is carried out by induction on k. The second stage of the rest of the proof is the conclusive one. At this stage we will establish the relationships of point (ii) for $k = k_*$: the relationships of point (iii) for the same k are then obtained automatically, and the relationships of point (i) are only related to the case $k \le k_* - 1$. At this conclusive stage, all the previously established relationships of points (i) - (iii) for $k \le k_* - 1$ will be used.

Now, we proceed with the continuation of the proof of Proposition 3.1.1.

The first stage. The derivation of validity of the relationships of point (ii) for any integer $1 \le k \le k_* - 1$ by induction on k.

The induction base is: $k = 1$. First, let us prove the relationships of (ii1). Recall that $n \ge 2 > k = 1$. Applying relationship (1.1.2), Lemma 1.1.1 with $u = \theta y$ and $\ell = k_* + 1$, and Lemma 1.1.3 we obtain the that

$$\mathbf{P}\{S_n > y\} = n \cdot \mathbf{P}\{S_n > y, X_n > \theta y\} \tag{3.3.2}$$

$$+ \sum_{m=2}^{n} (-1)^{m+1} \cdot \binom{n}{m} \cdot \mathbf{P}\{S_n > y, X_{n-m+1} > \theta y, \ldots, X_n > \theta y\} + \Delta_{2,1}(n,y).$$

Note that Lemma 1.1.3 implies that for any integer $n \ge 2$ and for $y \ge K_1 \cdot n^{1/\alpha_1}$,

$$|\Delta_{2,1}(n,y)| \le K_2 \cdot (n \cdot y^{-\alpha_1})^{k_*+1}.$$

By definition, the first term on the right-hand side of (3.3.2) is equal to $n \cdot P_1$. Approximating this term by the use of Lemma 3.1.8 we get that

$$n \cdot \mathbf{P}\{S_n > y, X_n > \theta y\} = n \cdot \Big(\sum_{l=1}^{\ell} c_{\alpha_l} \cdot (\theta y)^{-\alpha_l} - \mathbf{H}_1(y,0) + \mathbf{H}_1(y,0)$$

(3.3.3)

$$\cdot \int_{(1-\theta)y}^{\infty} z^0 \cdot d\big(-\mathbf{P}\{S_{n-1} > z\}\big) - \int_{-\infty}^{0} \Delta\mathbf{H}_1(y,z,0) \cdot d\mathbf{P}\{S_{n-1} < z\}$$

$$- \int_{0}^{(1-\theta)y} \Delta\mathbf{H}_1(y,z,0) \cdot d\big(-\mathbf{P}\{S_{n-1} > z\}\big)\Big) + \Delta_{3,1}(n,y).$$

Applying (3.1.33), we get that the absolute value of the remainder $\Delta_{3,1}(n,y)$ does not exceed

$$K_2 \cdot n \cdot \sup_{x \geq \theta y} |\Delta_+(x)|.$$

Now, recall that for $z \leq (1-\theta)y$,

$$\mathbf{H}_1(y,z) = \sum_{l=1}^{\ell} c_{\alpha_l} \cdot \big\{(\theta y)^{-\alpha_l} - (y-z)^{-\alpha_l}\big\}, \tag{3.3.4}$$

and that

$$\sum_{m=2}^{n} (-1)^{m+1} \cdot \binom{n}{m} \cdot \mathbf{P}\{S_n > y, X_{n-m+1} > \theta y, \ldots, X_n > \theta y\} = \Delta_{1,1}(n,y)$$

by definition. Combining these relationships with (3.3.2) and (3.3.3) we obtain that

$$\mathbf{P}\{S_n > y\} - n \cdot c_{\alpha_1} \cdot y^{-\alpha_1} = n \cdot \Big\{ \int_{-\infty}^{0} \Delta\mathbf{H}_1(y,z,0) \cdot d\mathbf{P}\{S_{n-1} < z\}$$

(3.3.5)

$$+ \int_{0}^{(1-\theta)y} \Delta\mathbf{H}_1(y,z,0) \cdot d\big(-\mathbf{P}\{S_{n-1} > z\}\big) \Big\} + n \cdot \mathbf{H}_1(y,0) \cdot \int_{(1-\theta)y}^{\infty} z^0 \cdot d\big(-\mathbf{P}\{S_{n-1} > z\}\big)$$

$$+ n \cdot \sum_{l=2}^{\ell} c_{\alpha_l} \cdot y^{-\alpha_l} + \Delta_{1,1}(n,y) + \Delta_{2,1}(n,y) + \Delta_{3,1}(n,y).$$

By definition, the initial two terms on the right-hand side of (3.3.5) are equal to the initial two terms on the right-hand side of (3.1.40) for the currently considered value $k = 1$. Now, due to the fact that for any integer $2 \leq i \leq \ell$, we have $\alpha_i \leq r < (k_* + 1) \cdot \alpha_1$, we obtain the result that the third term on the right-hand side of (3.3.5) can be rewritten as follows:

$$(3.3.5') \qquad n\cdot\sum_{l=2}^{t} c_{\alpha_l}\cdot y^{-\alpha_l} = \sum_{\substack{2\le l\le t\\ \alpha_l\le(k+1)\alpha_1}} n\cdot c_{\alpha_l}\cdot y^{-\alpha_l}.$$

Applying relationship (3.1.38′) for the case $k + 1 = 1$ and the equality

$$\theta^{-\alpha_l} - g_l(0) = y^{-\alpha_l}$$

we obtain the result that the right-hand side of (3.3.5′) coincides with $\Sigma_+^{(1)}(k + 1,n,y)$.

Moreover, it is not difficult to demonstrate that

$$n\cdot c_{\alpha_1}\cdot y^{-\alpha_1} = \Sigma_+^{(1)}(n,y),$$

and hence the left-hand side of (3.3.5) coincides with $\Delta_+(n,y,1)$. Therefore, (3.3.5) yields that

$$\Delta_+(n,y,1) = n\cdot(I_1(1,n\text{-}1,0) + I_2(1,n\text{-}1,0)) + n\cdot\mathbf{H}_1(y,0)\cdot I(0,1,n\text{-}1,0) +$$
$$\Sigma_+^{(1)}(k{+}1,n,y) + \Delta_{1,1}(n,y) + \Delta_{2,1}(n,y) + \Delta_{3,1}(n,y),$$

where all the remainders $\Delta_{1,1}(n,y)$, $\Delta_{2,1}(n,y)$, $\Delta_{3,1}(n,y)$ can be estimated with the required accuracy, which completes the proof of the induction base of (ii1). □

Now, we proceed with the proof of (ii2) of the induction base. It is required to establish the existence of a function $\mathbf{R}_{1,1}(\cdot)$ from $\mathbb{R}_+^1$ to $\mathbb{R}_+^1$ such that for any integer $n \ge 2$, for any real $\varepsilon > 0$, and for $y \ge \mathbf{R}_{1,1}(\varepsilon)n^{1/\alpha_1}$,

$$(3.3.6) \qquad \left| n\cdot\mathbf{H}_1(y,0)\cdot\int_{(1-\theta)y}^{\infty} z^0\cdot d\left(-\mathbf{P}\{S_{n-1}>z\}\right)\right| \le \epsilon\cdot n\cdot y^{-\alpha_1}.$$

It is easily seen that estimate (3.3.6) follows from (3.3.4) and Remark 1.1.1 to Proposition 1.1.1. The function $\mathbf{R}_{1,1}(\varepsilon)$ can be chosen to be equal to $K\cdot\varepsilon^{1/\alpha_1}$. □

Now, we proceed with the proof of (ii3) of the induction base. It is required to establish the existence of a function $\mathbf{R}_{2,1}(\cdot)$ from $\mathbb{R}_+^1$ to $\mathbb{R}_+^1$ such that for any integer $n \ge 2$, for any real $\varepsilon > 0$, and for $y \ge \mathbf{R}_{2,1}(\varepsilon)n^{1/\alpha_1}$,

$$(3.3.7) \qquad \left| n\cdot\int_{-\infty}^{0} \Delta\mathbf{H}_1(y,z,0)\cdot d\mathbf{P}\{S_{n-1}<z\}\right| \le \epsilon\cdot n\cdot y^{-\alpha_1};$$

$$(3.3.7') \qquad \left| n \cdot \int_0^{(1-\theta)y} \Delta \mathbf{H}_1(y,z,0) \cdot d\left(-\mathbf{P}\{S_{n-1} > z\}\right) \right| \le \epsilon \cdot n \cdot y^{-\alpha_1}.$$

We prove here only (3.3.7), since the latter inequality is proved following along the same lines, but in a simpler manner. Let us split the integral on the left-hand side of (3.3.7) into the sum of the integrals from $-\infty$ to $-y$ and from $-y$ to zero. Using the explicit representation for the function $\mathbf{H}_1(\cdot,\cdot)$ (cf. (3.3.4)) we obtain the following estimates:

$$\left| n \cdot \int_{-\infty}^{-y} \Delta \mathbf{H}_1(y,z,0) \cdot d\mathbf{P}\{S_{n-1} < z\} \right|$$

$$\le n \cdot \sum_{i=1}^{\ell} |c_{\alpha_i}| \cdot \int_{-\infty}^{-y} \left(y^{-\alpha_i} - (y-z)^{-\alpha_i}\right) \cdot d\mathbf{P}\{S_{n-1} < z\}$$

$$\le K \cdot n \cdot y^{-\alpha_1} \cdot \mathbf{P}\{S_{n-1} < -y\}.$$

Combining these inequalities with Remark 1.1.1 to Proposition 1.1.1 we derive the result that there exist positive constants K_1 and K_2 such that for any integer $n \ge 2$, and for $y \ge K_1 \, \epsilon^{-1/\alpha_1} \cdot n^{1/\alpha_1}$,

$$(3.3.8) \qquad \left| n \cdot \int_{-\infty}^{-y} \Delta \mathbf{H}_1(y,z,0) \cdot d\mathbf{P}\{S_{n-1} < z\} \right| \le K_2 \cdot \epsilon \cdot n \cdot y^{-\alpha_1}.$$

To estimate the integral from $-y$ to zero we again apply formula (3.3.4):

$$(3.3.9) \qquad \begin{aligned} &\left| n \cdot \int_{-y}^{0} \Delta \mathbf{H}_1(y,z,0) \cdot d\mathbf{P}\{S_{n-1} < z\} \right| \\ &\le n \cdot \sum_{i=1}^{\ell} |c_{\alpha_i}| \cdot \int_{-y}^{0} \left(y^{-\alpha_i} - (y-z)^{-\alpha_i}\right) \cdot d\mathbf{P}\{S_{n-1} < z\}. \end{aligned}$$

Let us evaluate an arbitrary i^{th} integral on the right-hand side of (3.3.9). For this purpose, we make the following change of variables $t = z/y$; we obtain that for any integer $1 \le i \le \ell$, i^{th} integral does not exceed

$$y^{-\alpha_1}\cdot\int_{-1}^{0}\left(1-(1-t)^{-\alpha_1}\right)\cdot d\mathbf{P}\{S_{n-1}<yt\}.$$

We estimate this expression following along the same lines as the rightmost expression on the right-hand side of (1.1.12). In particular, one should set

$$\phi(n,y) := y\cdot\left(y/n^{1/\alpha_1}\right)^{-\alpha_1/(1+\alpha_1)}.$$

Hence, for any integer $1 \le i \le \ell$,

$$n\cdot\int_{-y}^{0}\left(y^{-\alpha_i}-(y-z)^{-\alpha_i}\right)\cdot d\mathbf{P}\{S_{n-1}<z\} \le K\cdot\left(n\cdot y^{-\alpha_1}\right)^{1+1/(1+\alpha_1)}.$$

Combining this inequality with (3.3.8) and (3.3.9) we obtain the validity of inequality (3.3.7). □

The induction step. Note that in the case $k_* = 2$, after establishing the relationships of point (ii) for $k = 1 = k_* - 1$ (the induction base) we should immediately proceed with the conclusive stage of the proof. Hence, let us assume that $k_* \ge 3$. Let $k \ge 1$ be an arbitrary integer; $k \le k_* - 2$, and all the relationships of points (i) - (iii) for any integer $1 \le p \le$ k be fulfilled. Under these assumptions, let us show that all the relationships of point (ii) are also fulfilled for $p = k + 1 \le k_* - 1$.

Proof of (ii2) of the induction step. It suffices to establish the existence of a function $\mathbf{R}_{1,k+1}(\cdot)$ from $\mathbf{R}_+^1$ to $\mathbf{R}_+^1$ such that for any integer $n \ge 2$, for any integer $1 \le m \le (k+1) \wedge (n-1)$, for any non-negative integer $s \le [(k+1-m)\cdot\alpha_1]$, for any real $\varepsilon > 0$, and for $y \ge \mathbf{R}_{1,k+1}(\varepsilon)\cdot n^{1/\alpha_1}$,

$$\text{(3.3.10)}\qquad \left|\binom{n}{m}\cdot\frac{\mathbf{H}_m^{(s)}(y,0)}{s!}\cdot I(s,m,n-m,k+1-m)\right| \le \varepsilon\cdot\left(n\cdot y^{-\alpha_1}\right)^{k+1}.$$

Obviously, $0 \le (k+1-m)\cdot\alpha_1 \le k\cdot\alpha_1$, and $n - m \ge 1$. In the case $n - m = 1$ we apply expansion (0.5). The latter yields that there exists a function $X(\cdot)$ from $\mathbf{R}_+^1$ to $\mathbf{R}_+^1$ such that for any real $\varepsilon > 0$ and for $z \ge X(\varepsilon)$,

$$\left|F_{S_1}(z) - 1 + \sum_{i=1}^{t} c_{\alpha_i} \cdot z^{-\alpha_i}\right| \le \epsilon \cdot z^{-r}.$$

Therefore, there exists a function $\mathbf{V}(\cdot,\cdot)$ from $\mathbf{N} \otimes \mathbf{R}_+^1$ to $\mathbf{R}_+^1$ such that for any integer $k \le k_* - 1$, for any integer $1 \le m \le k + 1$, and for any real $\varepsilon > 0$, and for $z \ge \mathbf{V}(k,\varepsilon)$,

$$(3.3.11) \quad \begin{aligned} &|\Delta_+(1,z,k+1-m)| \\ &\le \left|F_{S_1}(z) - 1 + \sum_{i:\ \alpha_i \le (k+1-m)\alpha_1}^{t} c_{\alpha_i} \cdot z^{-\alpha_i}\right| \le \epsilon \cdot z^{-(k+1-m)\alpha_1}. \end{aligned}$$

If $n - m \ge 2$ then the already established inequalities of point (i) for any integer $1 \le p \le k$ imply that for any integer $1 \le m \le k \wedge (n\text{-}2)$, for any real $\varepsilon > 0$, and for $z \ge (\mathbf{L}_1(\varepsilon) \vee \dots \vee \mathbf{L}_k(\varepsilon))n^{1/\alpha_1}$,

$$(3.3.11') \quad |\Delta_+(n-m,z,k+1-m) \le \epsilon \cdot \left((n-m) \cdot z^{-\alpha_1}\right)^{k+1-m}.$$

On the other hand, if $m = k + 1$ then Remark 1.1.1 to Proposition 1.1.1 yields that for any integer $n > k + 1$, for any real $\varepsilon > 0$, and for $z \ge K\varepsilon^{-1/\alpha_1} \cdot n^{1/\alpha_1}$,

$$(3.3.11'') \quad |\Delta_+(n-k-1,z,0)| \le \epsilon.$$

The proof of inequality (3.3.10) is carried out by the use of integration by parts with the subsequent application of estimates (3.3.11) in the case n - m = 1; (3.3.11′) in the case $n - m \ge 2$ and $m \le k$, (3.3.11′′) in the case n - m ≥ 2 and $m = k + 1$.

Recall that

$$\mathrm{I}(s,m,n-m,k+1-m) = \int_{(1-m\theta)y}^{\infty} z^s \cdot d\left(-\Delta_+(n-m,z,k+1-m)\right)$$

(cf. point (ii1) of Proposition 3.1.1). Integrating by parts we obtain that

$$(3.3.12) \quad \int_{(1-m\theta)y}^{\infty} z^s \cdot d\left(-\Delta_+(n-m,z,k+1-m)\right)$$

$$= \left((1-m\theta)\cdot y\right)^{s}\cdot\Delta_{+}\left(n-m,(1-m\theta)y,k+1-m\right)+s\cdot\int\limits_{(1-m\theta)y}^{\infty}\Delta_{+}(n-m,z,k+1-m)\cdot dz,$$

where in the case $s = 0$, the integral on the right-hand side of (3.3.12) is set to be zero. Note that for any non-negative integer $m \le k_{*}$, we have $1/(1-m\theta) \le 2$. Hence, in order to apply inequalities (3.3.11) - (3.3.11′′) for estimating the right-hand side of (3.3.12) we should require that $y \ge 2\cdot\mathbf{Y}_k(\varepsilon)\,n^{1/\alpha_1}$, where

$$\mathbf{Y}_k(\epsilon) := \mathbf{L}_1(\epsilon)\vee\ldots\vee\mathbf{L}_k(\epsilon)\vee K\cdot\epsilon^{-1/\alpha_1}\vee\mathbf{V}(k,\epsilon).$$

In order to estimate the second factor on the left-hand side of (3.3.10), we apply inequality (3.1.22) from Lemma 3.1.5. We get that the left-hand side of (3.3.10) does not exceed

$$K\cdot n^{m}\cdot y^{-m\alpha_1-s}\cdot\left((1-m\theta)^{s-(k+1-m)\alpha_1}\cdot\epsilon\cdot y^{s}\cdot(ny^{-\alpha_1})^{k+1-m}\right.$$

$$\left.+\,s\cdot\epsilon\cdot n^{k+1-m}\cdot\int\limits_{(1-m\theta)y}^{\infty} z^{s-1-(k+1-m)\alpha_1}\cdot dz\right) \le K_2\cdot\epsilon\cdot(ny^{-\alpha_1})^{k+1},$$

which implies the validity of (ii2) of the induction step. □

Proof of (ii3) of the induction step. It is required to establish the existence of a function $\mathbf{R}_{2,k+1}(\cdot)$ from $\mathbf{R}_{+}^{1}$ to $\mathbf{R}_{+}^{1}$ such that for any integer $n \ge 2$, for any integer $1 \le m \le (k+1)\wedge(n-1)$, for any real $\varepsilon > 0$, and for $y \ge \mathbf{R}_{2,k+1}(\varepsilon)\,n^{1/\alpha_1}$,

$$\text{(3.3.13)}\qquad \left|\binom{n}{m}\cdot\mathrm{I}_1(m,n-m,k+1-m)\right| \le \epsilon\cdot\left(n\cdot y^{-\alpha_1}\right)^{k+1};$$

$$\text{(3.3.13′)}\qquad \left|\binom{n}{m}\cdot\mathrm{I}_2(m,n-m,k+1-m)\right| \le \epsilon\cdot\left(n\cdot y^{-\alpha_1}\right)^{k+1}.$$

First, we derive inequality (3.3.13′). To do this, we integrate the integral $\mathrm{I}_2(m,n-m,k+1-m)$ by parts:

$$\text{(3.3.14)}\qquad \begin{aligned} I_2(m,n-m,k+1-m) &= -\Delta\mathbf{H}_m\left(y,z,[(k+1-m)\cdot\alpha_1]\right) \\ &\cdot\Delta_{+}(n-m,z,k+1-m)\Big|_0^{(1-m\theta)y} + \int\limits_0^{(1-m\theta)y}\Delta_{+}(n-m,z,k+1-m) \end{aligned}$$

$$\cdot d\Delta\mathbf{H}_m\big(y,z,[(k+1-m)\cdot\alpha_1]\big).$$

Let us estimate the absolute value of the term outside the integral. It is easily seen that the absolute value of its first factor is estimated by the use of the Taylor expansion with Lagrange's form for the remainder of the function $\Delta\mathbf{H}_m(y,z,[(k+1-m)\cdot\alpha_1])$ (taken in both the upper and the lower limits). Subsequently, we apply Lemma 3.1.5 for the estimation of the absolute value of $([(k+1-m)\cdot\alpha_1])^{th}$ partial derivative in z of the function $\mathbf{H}_m$:

(3.3.15)
$$\big|\Delta\mathbf{H}_m\big(y,z,[(k+1-m)\cdot\alpha_1]\big)\big| = \frac{\mathbf{H}_m^{([(k+1-m)\cdot\alpha_1]+1)}(y,v)}{\big([(k+1-m)\cdot\alpha_1]+1\big)!}\cdot|z|^{[(k+1-m)\cdot\alpha_1]+1}$$
$$\le K\cdot y^{-m\alpha_1-[(k+1-m)\cdot\alpha_1]-1}\cdot|z|^{[(k+1-m)\cdot\alpha_1]+1},$$

where $|v| \le |z|$.

It follows from (3.3.15) and the definition of the function Δ_+ above the formulation of Proposition 3.1.1 that the absolute value of the term outside the integral taken in the lower limit does not exceed

(3.3.16)
$$\lim_{z\downarrow 0} K\cdot y^{-m\alpha_1-[(k+1-m)\alpha_1]-1}\cdot|z|^{[(k+1-m)\alpha_1]+1},$$
$$\cdot\Big(1+\sum_{j:\ \beta_j\le(k+1-m)\alpha_1}\big|c_{\beta_j}(n-m)\big|\cdot z^{-\beta_j}\Big) = 0.$$

Now, combining relationships (3.3.15), and (3.3.11) - (3.3.11′′) we obtain that for $y \ge 2\cdot\mathbf{Y}_k(\varepsilon)n^{1/\alpha_1}$, the absolute value of the term outside the integral taken in the upper limit does not exceed

(3.3.16′)
$$K_1\cdot y^{-m\alpha_1}\cdot|\Delta_+(n-m,(1-m\theta)y,k+1-m)|$$
$$\le K_2\cdot y^{-m\alpha_1}\cdot\epsilon\cdot\big((n-m)\cdot y^{-\alpha_1}\big)^{k+1-m}.$$

Combining (3.3.16) and (3.3.16′) we obtain the result that the absolute value of the first term on the right-hand side of (3.3.14) does not exceed expression (3.3.17):

(3.3.17) $$K_2\cdot\epsilon\cdot n^{k+1-m}\cdot y^{-(k+1)\alpha_1}.$$

Now, let us derive an analogous estimate for the absolute value of the integral on the right-hand side of (3.3.14). For this, we use the infinite differentiability of the function $\mathbf{H}_m$ in the second argument as well as the Taylor expansion with Lagrange's form for the remainder. We obtain the result that for $|z| \le (1-m\theta)\cdot y$,

$$\left|\frac{d}{dz}\Delta\mathbf{H}_m\big(y,z,[(k+1-m)\cdot\alpha_1]\big)\right| \le K\cdot y^{-m\alpha_1-1-[(k+1-m)\alpha_1]-1}$$

$$\cdot\ |z|^{[(k+1-m)\cdot\alpha_1]}.$$

Hence, the absolute value of the integral on the right-hand side of (3.3.14) does not exceed

$$K\cdot y^{-m\alpha_1-1-[(k+1-m)\alpha_1]}$$

(3.3.18)

$$\cdot\int_0^{(1-m\theta)y} |z|^{[(k+1-m)\alpha_1]}\,|\Delta_+(n-m,z,k+1-m)|\cdot dz.$$

The latter integral should be approached by different methods for large and for small z. To this end, let us split the integral into two parts: from zero to $\mathbf{Y}_k(\varepsilon)\cdot n^{1/\alpha_1}$ (I_1) and from $\mathbf{Y}_k(\varepsilon)\cdot n^{1/\alpha_1}$ to $(1-m\theta)\cdot y$ (I_2). Recall that $y \ge 2\cdot\mathbf{Y}_k(\varepsilon)\cdot n^{1/\alpha_1}$, and $1 - m\cdot\theta > 1/2$ for any non-negative integer $m \le k_*$. Hence, $(1-m\theta)\cdot y \ge \mathbf{Y}_k(\varepsilon)\cdot n^{1/\alpha_1}$.

We proceed with the estimation of $|\mathrm{I}_1|$. It follows from the definition of the function $\Delta_+(n,y,k)$ that

$$|\mathrm{I}_1| \le \int_0^{\mathbf{Y}_k(\epsilon)\cdot n^{1/\alpha_1}} \Big(1+\sum_{j:\ \beta_j\le(k+1-m)\alpha_1}|c_{\beta_j}(n-m)|\cdot z^{-\beta_j}\Big)\cdot|z|^{[(k+1-m)\alpha_1]}\cdot dz.$$

Recall that $c_{\beta_j}(n)$ is a polynomial on n whose power does not exceed $[\beta_j/\alpha_1]$. Therefore,

$$|\mathrm{I}_1| \le K\cdot|z|^{[(k+1-m)\alpha_1]+1}\cdot\Big(1+\sum_{j:\ \beta_j\le(k+1-m)\alpha_1} n^{[\beta_j/\alpha_1]}\cdot z^{-\beta_j}\Big)\Big|_{z=\mathbf{Y}_k(\epsilon)\cdot n^{1/\alpha_1}}.$$

Since the set $\{\beta_j\}$ consists of a finite number of elements we obtain that

$$K\cdot y^{-m\alpha_1-1-[(k+1-m)\alpha_1]}\cdot|\mathrm{I}_1| \le K_1\cdot y^{-m\alpha_1-1-[(k+1-m)\alpha_1]}\cdot|\mathrm{I}_1|$$

$$\cdot\max_{0\le\gamma\le(k+-m)\alpha_1}\Big(n^{[\gamma/\alpha_1]}\cdot\big(\mathbf{Y}_k(\epsilon)\cdot n^{1/\alpha_1}\big)^{[(k+1-m)\alpha_1]+1-\gamma}\Big)$$

(3.3.19)

$$\le K_1\cdot y^{-m\alpha_1-1-[(k+1-m)\alpha_1]}\cdot\big(\mathbf{Y}_k(\epsilon)\cdot n^{1/\alpha_1}\big)^{[(k+1-m)\alpha_1]+1}$$

$$= K_1 \cdot y^{-m\alpha_1} \cdot \left(y^{-1} \cdot Y_k(\epsilon) \cdot n^{1/\alpha_1}\right)^{[(k+1-m)\alpha_1]+1}.$$

Now, let us represent the latter factor from the rightmost expression of inequalities (3.3.19) as the following product:

$$\left(\frac{n^{1/\alpha_1}}{y}\right)^{(k+1-m)\alpha_1} \cdot \left(\frac{n^{1/\alpha_1}}{y}\right)^{1-\{(k+1-m)\alpha_1\}} \cdot Y_k(\epsilon)^{[(k+1-m)\alpha_1]+1}$$

Set

$$\kappa(k,m) := \frac{[(k+1-m)\alpha_1]+1}{1-\{(k+1-m)\alpha_1\}},$$

and

$$Z_k(\epsilon) := Y_k(\epsilon)^{\kappa(k,m)} \cdot \epsilon^{-1/(1-\{(k+1-m)\alpha_1\})}.$$

Then it is easily seen that for $y \geq Z_k(\epsilon) n^{1/\alpha_1}$,

$$\left(\frac{n^{1/\alpha_1}}{y}\right)^{1-\{(k+1-m)\alpha_1\}} \cdot Y_k(\epsilon)^{[(k+1-m)\alpha_1]+1} \leq \epsilon.$$

Hence, for $y \geq Z_k(\epsilon) n^{1/\alpha_1}$, the right-hand side (and therefore the left-hand side) of (3.3.19) do not exceed

$$K_1 \cdot y^{-m\alpha_1} \cdot \epsilon \cdot \left(n \cdot y^{-\alpha_1}\right)^{k+1-m}. \tag{3.3.20}$$

Now, let us estimate $|I_2|$. It follows from (3.3.11), (3.3.11′), and (3.3.11′′) that for any z from the area of integration,

$$|\Delta_+(n-m, z, k+1-m)| \leq \epsilon \cdot \left(n \cdot z^{-\alpha_1}\right)^{k+1-m}.$$

Hence,

$$K \cdot y^{-m\alpha_1 - 1 - [(k+1-m)\alpha_1]} \cdot |I_2| \leq K \cdot \epsilon \cdot n^{k+1-m} \cdot y^{-m\alpha_1 - 1 - [(k+1-m)\alpha_1]} \tag{3.3.21}$$

$$\cdot \int_{Y_k(\epsilon) \cdot n^{1/\alpha_1}}^{(1-m\theta) \cdot y} z^{-\{(k+1-m)\alpha_1\}} \cdot dz \leq K_1 \cdot \epsilon \cdot n^{k+1-m} \cdot y^{-(k+1)\alpha_1}.$$

Combining (3.3.18) - (3.3.21) we obtain the required estimate for the absolute value of the integral on the right-hand side of (3.3.14), which in turn (being combined with (3.3.14) and (3.3.17)) yields the validity of inequality (3.3.13′).

Now, let us establish the validity of inequality (3.3.13). To this end, we split the integral $I_1(m,n-m,k+1-m)$ on its left-hand side into the sum of integrals from $-\infty$ to $-(1-m\theta)\cdot y$ (I_3) and from $-(1-m\theta)\cdot y$ to zero (I_4). The integral I_4 is estimated by analogy with the integral $I_2(m,n-m,k+1-m)$:

$$\left|\binom{n}{m}\cdot I_4\right| \le \epsilon\cdot\left(n\cdot y^{-\alpha_1}\right)^{k+1}. \tag{3.3.22}$$

Integrating I_3 by parts we obtain that

$$\begin{aligned} I_3 &= \Delta\mathbf{H}_m\left(y,z,[(k+1-m)\alpha_1]\right)\cdot\Delta_-(n-m,-z,k+1-m)\Big|_{-\infty}^{-(1-m\theta)\cdot y} \\ &\quad - \int_{-\infty}^{-(1-m\theta)\cdot y}\Delta_-(n-m,-z,k+1-m)\cdot d\Delta\mathbf{H}_m\left(y,z,[(k+1-m)\alpha_1]\right). \end{aligned} \tag{3.3.23}$$

We proceed with the estimation of the term outside the integral. Applying Lemma 3.1.5 we obtain the result that the absolute value of its first factor does not exceed the following sum for $z \le -(1-m\theta)\cdot y$:

$$\sum_{s=0}^{[(k+1-m)\alpha_1]}\frac{\mathbf{H}_m^{(s)}(y,0)}{s!}\cdot|z|^s + |\mathbf{H}_m(y,z)| \le K\cdot y^{-m\alpha_1-1-[(k+1-m)\alpha_1]}\cdot|z|^{[(k+1-m)\alpha_1]}. \tag{3.3.24}$$

On the other hand, the absolute value of the second factor of the term outside the integral is easily estimated by the use of inequality (3.1.39′) of the induction hypothesis. We obtain that the absolute value of this factor does not exceed the following expression for $z \le -\mathbf{L}_{k+1-m}(\varepsilon)\cdot n^{1/\alpha_1}$:

$$\epsilon\cdot\left((n-m)\cdot|z|^{-\alpha_1}\right)^{k+1-m}. \tag{3.3.25}$$

Combining (3.3.24) - (3.3.25) we obtain the result that for $y \ge \mathbf{L}_{k+1-m}(\varepsilon)\cdot n^{1/\alpha_1}$, the absolute value of the term outside the integral on the right-hand side of (3.3.23) does not exceed

$$K_1\cdot\epsilon\cdot n^{k+1-m}\cdot y^{-(k+1)\alpha_1}. \tag{3.3.26}$$

Now, let us estimate the absolute value of the integral on the right-hand side of (3.3.23). Using the fact that the function $\mathbf{H}_m(\cdot,\cdot)$ is infinitely differentiable in the second argument we obtain that the absolute value of this integral does not exceed

$$(3.3.27)\qquad \int_{-\infty}^{-(1-m\theta)\cdot y} |\Delta_-(n-m,-z,k+1-m)| \cdot \left|\frac{d}{dz}\Delta\mathbf{H}_m(y,z,[(k+1-m)\cdot\alpha_1])\right| \cdot dz.$$

By analogy with (3.3.24) we obtain that for $z \le -(1-m\theta)y$,

$$(3.3.28)\qquad \begin{aligned} \left|\frac{d}{dz}\Delta\mathbf{H}_m(y,z,[(k+1-m)\cdot\alpha_1])\right| &\le \sum_{s=0}^{[(k+1-m)\cdot\alpha_1]} \frac{\mathbf{H}_m^{(s+1)}(y,0)}{s!}\cdot|z|^s \\ &+ \left|\mathbf{H}_m^{(1)}(y,z)\right| \le K\cdot y^{-m\alpha_1-[(k+1-m)\alpha_1]}\cdot|z|^{[(k+1-m)\alpha_1]-1}. \end{aligned}$$

Combining (3.3.25) and (3.3.28) we get that for $y \ge 2\cdot Y_k(\varepsilon)\cdot n^{1/\alpha_1}$, the integral (3.3.27) does not exceed

$$(3.3.29)\qquad \begin{aligned} &K\cdot\epsilon\cdot n^{k+1-m}\cdot y^{-m\alpha_1-[(k+1-m)\alpha_1]}\cdot\int_{-\infty}^{-(1-m\theta)\cdot y} z^{-1-\{(k+1-m)\cdot\alpha_1\}}\cdot dz \\ &\le \frac{K\cdot(1-m\theta)^{-\{(k+1-m)\cdot\alpha_1\}}}{\{(k+1-m)\cdot\alpha_1\}}\cdot\epsilon\cdot n^{k+1-m}\cdot y^{-(k+1)\alpha_1} \end{aligned}$$

(Recall that the denominator of the latter fraction is not equal to zero by Condition (3.1.6).)

Now, combining (3.3.23), (3.3.26), (3.3.27), and (3.3.29) we derive the required estimate for $|I_3|$. It only remains to note that the derived estimate for $|I_3|$ along with (3.3.22) yields inequality (3.3.13), which completes the proof of (ii3) of the induction step. □

Proof of (ii1) of the induction step. It is required to demonstrate that for any integer $n \ge 2$ and for any non-negative integer $k \le k_* - 1$,

$$(3.3.30)\qquad \begin{aligned} \Delta_+(n,y,k+1) &= -\sum_{m=1}^{(k+1)\wedge(n-1)} (-1)^{m+1}\cdot\binom{n}{m}\cdot\left\{I_1(m,n-m,k+1-m) + I_2(m,n-m,k+1-m)\right\} \\ &+ \sum_{m=1}^{(k+1)\wedge(n-1)} (-1)^{m+1}\cdot\binom{n}{m}\cdot\sum_{s=0}^{[(k+1-m)\alpha_1]} \frac{\mathbf{H}_m^{(s)}(y,0)}{s!}\cdot I(s,m,n-m,k+1-m) \\ &+ \Sigma_+^{(k+1)}(k_*+1,n,y) + \Delta_{1,k+1}(n,y) + \Delta_{2,k+1}(n,y) + \Delta_{3,k+1}(n,y), \end{aligned}$$

where

$$\Delta_{1,k+1}(n,y) := \begin{cases} \sum_{m=k+2}^{n} (-1)^{m+1} \cdot \binom{n}{m} \cdot P_m & \text{if } k+1 < n; \\ 0 & \text{otherwise,} \end{cases}$$

and the remainders $\Delta_{2,k+1}(n,y)$ and $\Delta_{3,k+1}(n,y)$ are such that there exist positive constants K_1, K_2, and K_3 that could depend on k such that for any integer $n \geq 2$, for any integer $0 \leq k \leq k_* - 1$, and for $y \geq K_1 n^{1/\alpha_1}$,

$$|\Delta_{2,k+1}(n,y)| \leq K_2 \cdot \left(n \cdot y^{-\alpha_1}\right)^{k_*+1};$$

$$|\Delta_{3,k+1}(n,y)| \leq K_3 \cdot n \cdot \sup_{x \geq \theta y} |\Delta_+(x)|.$$

Proof of (3.3.30) is based on an application of relationship (3.1.40) and the recursive formulas (3.1.38) - (3.1.38′). Note that the proof in turn is split into several stages.

I. First, we replace in (3.1.40) the expression $\Sigma_+^{(k)}(k_*+1,n,y)$ by the following expression:

$$\Sigma_+^{(k+1)}(k_*+1,n,y) + \Sigma_+^{(k)}(k+1,n,y)$$

- the sum of the third and the fourth terms on the right-hand side of (3.1.38) with $k_1 := k_* + 1$ (this replacement is justified by formula (3.1.38′)).

Now, let us subtract the sum of the third and the fourth terms on the right-hand side of (3.1.38) and then carry over the term $\Sigma_+^{(k)}(k+1,n,y)$ from the right-hand side to the left-hand side. Applying (3.1.38) we obtain the result that the left-hand side turns into $\Delta_+(n,y,k+1)$, and the right-hand side turns into

$$- \sum_{m=1}^{k\wedge(n-1)} (-1)^{m+1} \cdot \binom{n}{m} \cdot \left\{ I_1(m,n-m,k-m) + I_2(m,n-m,k-m) \right\}$$

$$+ \sum_{m=1}^{k\wedge(n-1)} (-1)^{m+1} \cdot \binom{n}{m} \cdot \sum_{s=0}^{[(k-m)\alpha_1]} \frac{H_m^{(s)}(y,0)}{s!} \cdot I(s,m,n-m,k-m)$$

$$+ \Sigma_+^{(k+1)}(k_*+1,n,y) + \Delta_{1,k}(n,y) + \Delta_{2,k}(n,y) + \Delta_{3,k}(n,y)$$

- the sum of the third and the fourth terms on the right-hand sides of (3.1.38) and (3.1.38′).

Adding the terms with the same indices and transforming the obtained expressions by the use of Lemma 3.1.10 we get the following representation:

$$\Delta_+(n,y,k+1) = \sum_{m=1}^{k\wedge(n-1)} (-1)^{m+1}\cdot\binom{n}{m}\cdot\left\{I_1(m,n-m,k-m) + I_2(m,n-m,k-m)\right\}$$

(3.3.31)

$$+ \sum_{m=1}^{k\wedge(n-1)} (-1)^{m+1}\cdot\binom{n}{m}\cdot\sum_{s=0}^{[(k-m)\alpha_1]} \frac{\mathbf{H}_m^{(s)}(y,0)}{s!}\cdot I(s,m,n-m,k-m)$$

$$+ \Sigma_+^{(k+1)}(k_*+1,n,y) + \Delta_{1,k}(n,y) + \Delta_{2,k}(n,y) + \Delta_{3,k}(n,y),$$

$$- I_{\{k+1\le n\}}\cdot(-1)^{k+2}\cdot\binom{n}{k+1}\cdot\left\{\left(\sum_{i=1}^{\ell} c_{\alpha_i}\cdot(\theta y)^{-\alpha_i}\right)^{k+1} - \mathbf{H}_{k+1}(y,0)\right\}$$

$$- \sum_{m=1}^{k\wedge(n-1)} (-1)^{m+1}\cdot\binom{n}{m}\cdot\left\{\sum_{s=0}^{[(k-m)\alpha_1]} \frac{\mathbf{H}_m^{(s)}(y,0)}{s!}\cdot\sum_{j:\,(k-m)\alpha_1<\beta_j\le(k+1-m)\alpha_1} \frac{\beta_j}{\beta_j-s}\right.$$

$$\cdot(1-m\theta)^{s-\beta_j}\cdot c_{\beta_j}(n-m)\cdot y^{s-\beta_j} - \sum_{s=[(k-m)\alpha_1]+1}^{[(k+1-m)\alpha_1]} \frac{\mathbf{H}_m^{(s)}(y,0)}{s!}$$

$$\cdot\left(\sum_{j:\,s<\beta_j\le(k+1-m)\alpha_1} \frac{\beta_j}{\beta_j-s}\cdot(1-m\theta)^{s-\beta_j}\cdot c_{\beta_j}(n-m)\cdot y^{s-\beta_j} - b_s(S_{n-m})\right)$$

$$+ \sum_{j:\,(k-m)\alpha_1<\beta_j\le(k+1-m)\alpha_1} c_{\beta_j}(n-m)\cdot\int_0^{(1-m\theta)y} \Delta\mathbf{H}_m(y,z,[\beta_j])\cdot d\left(-z^{-\beta_j}\right)$$

$$\left.+ \sum_{j:\,(k-m)\alpha_1<\beta_j\le(k+1-m)\alpha_1} d_{\beta_j}(n-m)\cdot\int_{-\infty}^{0} \Delta\mathbf{H}_m(y,z,[\beta_j])\cdot d|z|^{-\beta_j}\right\}.$$

II. Now, we proceed with the transformation of the remainder $\Delta_{1,k}(n,y)$. Obviously,

$$\text{(3.3.32)} \qquad \Delta_{1,k}(n,y) = I_{\{k+1\le n\}}\cdot(-1)^{k+2}\cdot\binom{n}{k+1}\cdot P_{k+1} + \Delta_{1,k+1}(n,y).$$

We estimate the product $\binom{n}{k+1}\cdot P_{k+1}$ by applying Lemma 3.1.8 if $k + 1 < n$ or Lemma 3.1.8′ if $k + 1 = n$. Then (3.1.32) and (3.1.32) yield that

$$
\begin{aligned}
\binom{n}{k+1}\cdot P_{k+1} &= \binom{n}{k+1}\cdot\left\{\left(\sum_{i=1}^{l} c_{\alpha_i}\cdot(\theta y)^{-\alpha_i}\right)^{k+1} - \mathbf{H}_{k+1}(y,0)\right\} \\
&- \mathrm{I}_{\{k+2\le n\}}\cdot\binom{n}{k+1}\cdot\Big\{\mathrm{I}_1(k+1,n-k-1,0) + \mathrm{I}_2(k+1,n-k-1,0) \\
&- \mathbf{H}_{k+1}(y,0)\cdot\mathrm{I}(0,k+1,n-k-1,0)\Big\} + \Delta_{k+1}(n,y),
\end{aligned}
\tag{3.3.33}
$$

where the remainder $\Delta_{k+1}(\cdot,\cdot)$ is such that there exist positive constants K_1 and K_2 such that for any integer $n \ge 2$ and for $y \ge K_1 \cdot n^{1/\alpha_l}$,

$$
|\Delta_{k+1}(n,y)| \le K_2\cdot n\cdot \sup_{x\ge\theta y} |\Delta_+(x)|. \tag{3.3.34}
$$

Combining (3.3.32) and (3.3.33) we obtain that

$$
\begin{aligned}
&\Delta_{1,k}(n,y) - \mathrm{I}_{\{k+1\le n\}}\cdot(-1)^{k+2}\cdot\binom{n}{k+1}\cdot\left\{\left(\sum_{i=1}^{l} c_{\alpha_i}\cdot(\theta y)^{-\alpha_i}\right)^{k+1} - \mathbf{H}_{k+1}(y,0)\right\} \\
&- \mathrm{I}_{\{k+2\le n\}}\cdot(-1)^{k+1}\cdot\binom{n}{k+1}\cdot\Big\{\mathrm{I}_1(k+1,n-k-1,0) + \mathrm{I}_2(k+1,n-k-1,0) \\
&- \mathbf{H}_{k+1}(y,0)\cdot\mathrm{I}(0,k+1,n-k-1,0)\Big\} = \mathrm{I}_{\{k+1\le n\}}\cdot(-1)^{k+2}\cdot\Delta_{k+1}(n,y) + \Delta_{1,k+1}(n,y).
\end{aligned}
\tag{3.3.35}
$$

It is easily seen that in the case $k + 2 \le n$, we have $(k+1) \wedge (n-1) = k + 1$, and the third term on the left-hand side of (3.3.35) coincides with the sum of $(k+1)^{\text{th}}$ terms of the first and the second sums on the right-hand side of (3.3.30).

III. Now, let us transform the second sum on the right-hand side of (3.3.31). For this, we represent the integrals $\mathrm{I}(s,m,n-m,k-m)$ (contained in this sum) in the following form:

$$
\begin{aligned}
\mathrm{I}(s,m,n-m,k-m) &= \mathrm{I}(s,m,n-m,k+1-m) \\
&+ \int_{(1-m\theta)y}^{\infty} z^s\cdot d\big(\Delta_+(n-m,z,k+1-m) - \Delta_+(n-m,z,k-m)\big).
\end{aligned}
\tag{3.3.36}
$$

Obviously, the latter integral is equal to

$$+\int_{(1-m\theta)y}^{\infty} z^s \cdot d\Big(-\sum_{j:\,(k-m)\alpha_1<\beta_j\le(k+1-m)\alpha_1} c_{\beta_j}(n-m)\cdot z^{-\beta_j}\Big)$$

$$=\sum_{j:\,(k-m)\alpha_1<\beta_j\le(k+1-m)\alpha_1} \cdot\frac{\beta_j}{\beta_j-s}\cdot(1-m\theta)^{s-\beta_j}\cdot c_{\beta_j}(n-m)\cdot y^{s-\beta_j}.$$

It is easily seen that the latter integral converges due to the fact that $\beta_j > (k\text{-}m)\cdot\alpha_1$, and $s \le [(k\text{-}m)\cdot\alpha_1]$. Now, the above equality along with (3.3.36) yields that

$$\sum_{m=1}^{k\wedge(n-1)}(-1)^{m+1}\cdot\binom{n}{m}\cdot\sum_{s=0}^{[(k-m)\alpha_1]}\frac{\mathbf{H}_m^{(s)}(y,0)}{s!}\cdot I(s,m,n-m,k-m)$$

(3.3.37)
$$=\sum_{m=1}^{k\wedge(n-1)}(-1)^{m+1}\cdot\binom{n}{m}\cdot\sum_{s=0}^{[(k-m)\alpha_1]}\frac{\mathbf{H}_m^{(s)}(y,0)}{s!}\cdot I(s,m,n-m,k+1-m)$$
$$+\sum_{m=1}^{k\wedge(n-1)}(-1)^{m+1}\cdot\binom{n}{m}\cdot\sum_{s=0}^{[(k-m)\alpha_1]}\frac{\mathbf{H}_m^{(s)}(y,0)}{s!}$$
$$\cdot\sum_{j:\,(k-m)\alpha_1<\beta_j\le(k+1-m)\alpha_1}\frac{\beta_j}{\beta_j-s}\cdot(1-m\theta)^{s-\beta_j}\cdot c_{\beta_j}(n-m)\cdot y^{s-\beta_j}.$$

Note that the leftmost sum over m on the right-hand side of (3.3.37) coincides with the sum of the initial $k \wedge (n\text{-}1)$ terms of the second sum on the right-hand side of (3.3.30), for which the tails of the interior sums over s, corresponding to the summation over s such that $[(k\text{-}m)\alpha_1]+1 \le s \le [(k+1\text{-}m)\alpha_1]$, are dropped. These 'dropped tails' of the interior sums over s will be separated in the fourth stage. Let us also mention that the second sum on the right-hand side of (3.3.37) will cancel out with the first term within the brackets from the rightmost sum over m on the right-hand side of (3.3.31).

IV. Consider an arbitrary (m^{th}) term of the leftmost sum on the right-hand side of (3.3.31), where $1 \le m \le k \wedge (n\text{-}1)$. Let us distinguish the three different cases: $[(k+1\text{-}m)\alpha_1] = [(k\text{-}m)\alpha_1]$, $[(k+1\text{-}m)\alpha_1] = [(k\text{-}m)\alpha_1] + 1$, and $[(k+1\text{-}m)\alpha_1] = [(k\text{-}m)\alpha_1] + 2$. By this, we exhaust all the possible values of $[(k+1\text{-}m)\alpha_1]$, since if $\alpha_1 < 2$ then $0 \le [(k+1\text{-}m)\alpha_1] - [(k\text{-}m)\alpha_1] \le 2$.

IV.A (the simplest one): $[(k+1\text{-}m)\alpha_1] = [(k\text{-}m)\alpha_1]$. Then it is relatively easy to see that

$$\mathrm{I}_1(m,n-m,k-m) = \mathrm{I}_1(m,n-m,k+1-m)$$

(3.3.38)

$$+ \sum_{j:\,(k-m)\alpha_1<\beta_j\le(k+1-m)\alpha_1} d_{\beta_j}(n-m)\cdot\int_{-\infty}^{0}\Delta\mathbf{H}_m\big(y,z,[(k+1-m)\alpha_1]\big)\cdot d|z|^{-\beta_j},$$

and

$$\mathrm{I}_2(m,n-m,k-m) = \mathrm{I}_2(m,n-m,k+1-m)$$

(3.3.38′)

$$+ \sum_{j:\,(k-m)\alpha_1<\beta_j\le(k+1-m)\alpha_1} c_{\beta_j}(n-m)\cdot\int_{0}^{(1-m\theta)y}\Delta\mathbf{H}_m\big(y,z,[(k+1-m)\alpha_1]\big)\cdot d(-z^{-\beta_j}).$$

Let us point out that for the (currently) considered case, the sum of the leftmost terms on the right-hand sides of (3.3.38) and (3.3.38′) forms m^{th} term of the leftmost sum on the right-hand side of (3.3.30). Let us emphasize that the rightmost terms on the right-hand sides of (3.3.38) and (3.3.38′) will cancel out with the two latter terms within the brackets from the latter sum over m on the right-hand side of (3.3.31).

IV.B $[(k+1-m)\alpha_1] = [(k-m)\alpha_1] + 1$. We restrict ourselves by considering the integral $\mathrm{I}_2(m,n-m,k-m)$ only. After some algebra we obtain that

$$\mathrm{I}_2(m,n-m,k-m) = \sum_{j:\,(k-m)\alpha_1<\beta_j\le[(k+1-m)\alpha_1]+1} c_{\beta_j}(n-m)$$

(3.3.39)

$$\cdot\int_{0}^{(1-m\theta)y}\Delta\mathbf{H}_m\big(y,z,[(k-m)\alpha_1]\big)\cdot d\big(-z^{-\beta_j}\big)$$

$$+\int_{0}^{(1-m\theta)y}\Delta\mathbf{H}_m\big(y,z,[(k-m)\alpha_1]\big)\cdot d\big(-\Delta^{+}_{<[(k-m)\alpha_1]+1}(S_{n-m};z)\big).$$

Now, we represent the latter integral as the following sum:

$$\int_{0}^{(1-m\theta)y}\Delta\mathbf{H}_m\big(y,z,[(k-m)\alpha_1]\big)\cdot d\big(-\Delta^{+}_{<[(k-m)\alpha_1]+1}(S_{n-m};z)\big)$$

$$= \frac{\mathbf{H}_m^{([(k-m)\alpha_1]+1)}(y,0)}{([(k-m)\alpha_1]+1)!} \cdot \int_0^{(1-m\theta)y} z^{[(k-m)\alpha_1]+1} \cdot d\left(-\Delta^+_{<[(k-m)\alpha_1]+1}(S_{n-m};z)\right)$$

$$+ \int_0^{(1-m\theta)y} \Delta\mathbf{H}_m(y,z,[(k-m)\alpha_1]+1) \cdot d\left(-\Delta^+_{<[(k-m)\alpha_1]+1}(S_{n-m};z)\right).$$

Combining this representation with the equality $[(k+1-m)\alpha_1] = [(k-m)\alpha_1] + 1$ we obtain the following result:

$$\mathrm{I}_2(m,n-m,k-m) = \sum_{j:\ (k-m)\alpha_1<\beta_j<[(k+1-m)\alpha_1]+1} c_{\beta_j}(n-m)$$

$$(3.3.40) \quad \cdot \int_0^{(1-m\theta)y} \Delta\mathbf{H}_m(y,z,[(k-m)\alpha_1]) \cdot d\left(-z^{-\beta_j}\right)$$

$$+ \frac{\mathbf{H}_m^{([(k+1-m)\alpha_1]+1)}(y,0)}{([(k+1-m)\alpha_1]+1)!} \cdot \int_0^{(1-m\theta)y} z^{[(k+1-m)\alpha_1]+1} \cdot d\left(-\Delta^+_{<[(k+1-m)\alpha_1]+1}(S_{n-m};z)\right)$$

$$+ \int_0^{(1-m\theta)y} \Delta\mathbf{H}_m(y,z,[(k+1-m)\alpha_1]) \cdot d\left(-\Delta^+_{<[(k+1-m)\alpha_1]+1}(S_{n-m};z)\right).$$

Now, we represent the integral from zero to $(1-m\theta)y$ from the second term on the right-hand side of (3.3.40) as the difference of the integrals from zero to $(1-m\theta)y$ and from $(1-m\theta)y$ to infinity. The integral from $(1-m\theta)y$ to infinity in turn is represented in the following form:

$$\int_{(1-m\theta)y}^{\infty} z^{[(k+1-m)\alpha_1]} \cdot d\left(-\sum_{j:\ [(k+1-m)\alpha_1]<\beta_j\le(k+1-m)\alpha_1} c_{\beta_j}(n-m)\cdot z^{-\beta_j}\right)$$

$$+ \int_{(1-m\theta)y}^{\infty} z^{[(k+1-m)\alpha_1]} \cdot d\left(-\Delta_+(n-m,z,k+1-m)\right).$$

Note that the latter integral coincides with $\mathrm{I}([(k+1-m)\alpha_1],m,n-m,k+1-m)$. Hence,

$$\int_0^{(1-m\theta)y} z^{[(k+1-m)\alpha_1]} \cdot d\left(-\Delta^{+}_{<[(k+1-m)\alpha_1]}(S_{n-m};z)\right)$$

(3.3.41)

$$= \int_0^{\infty} z^{[(k+1-m)\alpha_1]} \cdot d\left(-\Delta^{+}_{<[(k+1-m)\alpha_1]}(S_{n-m};z)\right)$$

$$- \sum_{j:\ [(k+1-m)\alpha_1]<\beta_j\le(k+1-m)\alpha_1} c_{\beta_j}(n-m) \cdot \frac{\beta_j}{\beta_j - [(k+1-m)\alpha_1]}$$

$$\cdot \left((1-m\theta)y\right)^{[(k+1-m)\alpha_1]-\beta_j} - \mathrm{I}\left([(k+1-m)\alpha_1],m,n-m,k+1-m\right).$$

Now, let us transform the rightmost integral on the right-hand side of (3.3.40). After some algebraic operations we obtain the result that this integral is equal to the following sum:

(3.3.42)

$$\sum_{j:\ [(k+1-m)\alpha_1]<\beta_j\le(k+1-m)\alpha_1} c_{\beta_j}(n-m) \cdot \int_0^{(1-m\theta)y} \Delta\mathbf{H}_m\left(y,z,[(k+1-m)\alpha_1]\right)\cdot d\left(-z^{-\beta_j}\right)$$

$$+ \int_0^{(1-m\theta)y} \Delta\mathbf{H}_m\left(y,z,[(k+1-m)\alpha_1]\right)\cdot d\left(-\Delta_{+}(n-m,z,k+1-m)\right).$$

Note that the second integral is equal to $\mathrm{I}_2(m,n-m,k+1-m)$ by definition. Hence, combining (3.3.40) - (3.3.42) we obtain the following result:

(3.3.43)

$$\mathrm{I}_2(m,n-m,k-m) = \mathrm{I}_2(m,n-m,k+1-m)$$

$$- \sum_{s=[(k-m)\alpha_1]+1}^{[(k+1-m)\alpha_1]} \frac{\mathbf{H}_m^{(s)}(y,0)}{s!} \cdot \Bigg(\mathrm{I}(s,n-m,k+1-m) + \sum_{s<\beta_j\le(k+1-m)\alpha_1} c_{\beta_j}(n-m)$$

$$\cdot \frac{\beta_j}{\beta_j - s} \cdot (1-m\theta)^{s-\beta_j} \cdot y^{s-\beta_j} - \int_0^{\infty} z^{s} \cdot d\left(-\Delta^{+}_{<s}(S_{n-m};z)\right)\Bigg)$$

$$+ \sum_{j:\ (k-m)\alpha_1<\beta_j\le(k+1-m)\alpha_1} c_{\beta_j}(n-m) \cdot \int_0^{(1-m\theta)y} \Delta\mathbf{H}_m\left(y,z,[\beta_j]\right)\cdot d\left(-z^{-\beta_j}\right).$$

Note that for the currently considered case, all the sums over s consist of only one term, since $[(k+1-m)\alpha_1] = [(k-m)\alpha_1] + 1$. It also deserves to mention that we can unite the leftmost sum on the right-hand side of (3.3.40) and the leftmost sum from the expression (3.3.40) obtaining the rightmost sum on the right-hand side of (3.3.43) as the result. This is so because Condition (3.1.6), implies $\beta_j \neq [(k+1-m)\alpha_1]$. Let us emphasize that relationship (3.3.43) can be generalized in order to cover the case IV.A. To this end, we set all the sums over s to be zero if $[(k-m)\alpha_1] = [(k+1-m)\alpha_1]$ (This would turn (3.3.43) into (3.3.38′).) A similar (but simpler) relationship is also valid for $I_1(m,n-m,k-m)$:

$$\begin{aligned} \text{(3.3.43′)} \quad I_1(m,n-m,k-m) &= I_1(m,n-m,k+1-m) \\ &- \sum_{s=[(k-m)\alpha_1]+1}^{[(k+1-m)\alpha_1]} \frac{\mathbf{H}_m^{(s)}(y,0)}{s!} \cdot \int_{-\infty}^{0} z^s \cdot d\left(\Delta_{<s}^{-}(S_{n-m};|z|)\right) \\ &+ \sum_{j:\,(k-m)\alpha_1<\beta_j\le(k+1-m)\alpha_1} d_{\beta_j}(n-m)\cdot\int_{-\infty}^{0}\Delta\mathbf{H}_m(y,z,[\beta_j])\cdot d|z|^{-\beta_j}. \end{aligned}$$

Note that the sum of the leftmost terms on the right-hand sides of (3.3.43) and (3.3.43′) forms the mth term of the leftmost sum on the right-hand side of (3.3.30), the second sum on the right-hand side of (3.3.43) being combined with the interior sum from zero to $[(k-m)\alpha_1]$ from the mth term of the leftmost sum on the right-hand side of (3.3.37) forms the mth term of the second sum on the right-hand side of (3.3.30). In addition, the sum of the fourth term on the right-hand side of (3.3.43) and the second term on the right-hand side of (3.3.43′) is equal to

$$\text{(3.3.44)} \quad \sum_{s=[(k-m)\alpha_1]+1}^{[(k+1-m)\alpha_1]} b_s(S_{n-m}) \cdot \frac{\mathbf{H}_m^{(s)}(y,0)}{s!}$$

by (3.1.2), where $b_s(S_{n-m})$ stands for the sth pseudomoment in Cramér's sense of r.v. S_{n-m} (Note that the finiteness of this pseudomoment follows from the induction hypothesis.) Moreover, expression (3.3.44) cancels out with the third term within the brackets from the latter sum over m on the right-hand side of (3.3.31); the sum of the third term on the right-hand side of (3.3.43) and the rightmost terms on the right-hand sides of (3.3.43), and (3.3.43′) cancel out with the sum of the other terms within the brackets from the latter sum over m on the right-hand side of (3.3.31). This concludes the case IV.B.

IV.C: $[(k+1-m)\alpha_1] = [(k-m)\alpha_1] + 2$. Transforming the integrals $I_1(m,n-m,k-m)$ and $I_2(m,n-m,k-m)$ by analogy with the case IV.B enables one to demonstrate that relationships (3.3.43) - (3.3.43′) also remain valid in this case. Here, we drop the computations for the case $[(k+1-m)\alpha_1] = [(k-m)\alpha_1] + 2$ since they differ from those related to the case $[(k+1-m)\alpha_1] = [(k-m)\alpha_1] + 1$ only by their cumbersomeness.

V. Let us conclude the proof of point (i1) of the induction step. For this reason, we combine all the results obtained at the four previous stages. Recall that at the first stage, we have derived relationship (3.3.31) for $\Delta_+(n,y,k+1)$ starting from the induction hypothesis (i.e., relationship (3.1.40)). At the second stage, we have transformed the expression

$$\Delta_{1,k}(n,y) - I_{k+1\le n}\cdot(-1)^{k+2}\cdot\binom{n}{k+1}\cdot\left\{\left(\sum_{i=1}^{t} c_{\alpha_i}\cdot(\theta y)^{-\alpha_i}\right)^{k+1} - \mathbb{H}_{k+1}(y,0)\right\}$$

from the right-hand side of (3.3.31) representing it as the sum of the remainder $\Delta_{1,k+1}(n,y)$, another remainder (which is absorbed into $\Delta_{3,k+1}(n,y)$), and $(k+1)^{th}$ terms of the two leftmost sums on the right-hand side of (3.3.30) (Recall that in the case when these terms emerge in (3.3.30), the additional factor $I_{\{k+2\le n\}}$ is acquired, due to the equality $k + 1 = (k+1) \wedge (n-1)$.) At the third and the fourth stages, we have transformed the leftmost and next-to-leftmost sums on the right-hand side of (3.3.31) turning them into parts of the two leftmost sums on the right-hand side of (3.3.30), which correspond to the summation over m, $1 \le m \le k \wedge (n-1)$, and the sum of additional refining terms. These refining terms in turn are absorbed into the rightmost sum over m on the right-hand side of (3.3.31). Hence, the combination of the second, the third and the fourth stages enables one to establish the equality of the right-hand sides of relationships (3.3.30) and (3.3.31). This completes the proof of point (i1), and hence the proof of the induction step. □

Now, in order to complete the proof of Proposition 3.1.1, the relationships of point (i) should be established in the case $k = k_*$. The proof of this is the same as the just completed proof of the induction step. The only reason why the case $k = k_*$ is considered separately is its significance (since this is the case which is used in the proof of Theorem 3.1.1), and our will to unify the proofs of points (i)-(iii) carrying them out by induction on $k \le k_* - 1$ (Recall that all the assertions of point (i) are only related to such k.) □

Chapter 4

Limit Theorems on Large Deviations for Order Statistics.

First, let us give some notation. Throughout this chapter, we will denote the order statistics constructed by the sample $\{X_n, n \geq 1\}$ of i.i.d. random variables with common distribution function $F(\cdot)$ by $X_n(n) \leq \ldots < X_1(n)$.

Recall that in this chapter, we study the asymptotic behavior of the *probabilities of large deviations for order statistics and their sums*, i.e., the asymptotics of the probabilities such as $\mathbf{P}\{X_1(n) > y\}$, $\mathbf{P}\{X_n(n) > y\}$, $\mathbf{P}\{S_n - X_1(n) - \ldots - X_k(n) > y\}$, etc., where n, y and k vary such that these probabilities tend to zero.

Section 4.1 is mainly devoted to the investigation of the probabilities of large deviations for maxima $X_1(n)$ (a certain result for minima $X_n(n)$ is obtained as a corollary). The study of the asymptotic behavior of $\mathbf{P}\{S_n - X_1(n) - \ldots - X(n) > y\}$ is carried out in Section 4.2. The remaining Sections 4.3 and 4.4 are of illustrative character containing the results on the asymptotics of conditional probabilities both for maxima and for trimmed sums.

4.1 Large Deviations for Maxima: the Tail Approximation/Extreme Value Approximation Alternative.

Let us first consider the case when a given a priori random sample $\{X_n, n \geq 1\}$ has the standard normal distribution. Recall that in this case, the distribution of the properly centered and normalized maximum $X_1(n)$ converges weakly to the Gumbel distribution $\Lambda(x)$ (see formula (0.11) in the Introduction).

It is interesting to note that one can refine relationship (0.11′) by the use of formula (10) of Hall (1979) and Theorem 2 of Cohen (1982b) as follows:

$$\mathbf{P}\{X_1(n) \le A'_n + B'_n \cdot x\} - \Lambda(x)$$

(4.1.1)

$$= \Lambda(x)\cdot e^{-x}\cdot\frac{x^2/2+x+1}{2\cdot\log n} + o\left(\frac{1}{\log n}\right)$$

as $n \to \infty$ uniformly in $x \in \mathbf{R}_+^1$, where the centering and normalizing sequences A'_n and B'_n are defined below formula (0.11′). Note that formula (10) of Hall (1979) is in fact an auxiliary result of that work used for the derivation of (0.11′), whereas slightly different centering and normalizing sequences were chosen in Theorem 2 of Cohen (1982b). Let us also point out that Theorem 2 of Cohen (1982b) covers a wide class of distributions (which contains the normal distribution) known as class **N**.

It was already mentioned in the Introduction that the rate of convergence in (0.11) is very slow and worst on the tails (cf., e.g., Hall's remark quoted below formula (0.11′)). In this respect, the following result was derived in Hall (1980) Theorem 3, which can be viewed as a non-uniform estimate in (0.11) taking into account right-hand large deviations.

Let $a_n > 0$ be defined from the equation

$$2\pi \cdot a_n^2 \cdot \exp\{a_n^2\} = n^2,$$

and let

$$z_n(x) := (2\pi)^{-1/2} \cdot n \cdot x^{-1} \cdot \exp\{-x^2/2\}.$$

Then for $x \ge a_n$,

$$\exp\{-z_n(x)\cdot[1-x^{-2}+3x^{-4}+z_n(x)/2(n-1)]\}$$

(4.1.2) $$\le \mathbf{P}\{X_1(n) \le x\} \le \exp\{-z_n(x)\cdot[1-x^{-2}]\}.$$

Now, let us point out that Theorem 3 of de Haan and Hordijk (1972) (see also Remark on p. 1195 therein) implies that the asymptotics of the probabilities of right-hand large deviations for the centered and normalized maximum from the normal sample is given by the tail of function Λ only in a very narrow range of large deviations. Namely,

(4.1.3) $$\mathbf{P}\{X_1(n) > A_n + B_n \cdot x\} \sim 1 - \Lambda(x)$$

as $n \to \infty$ and $x \to \infty$, $x = o((\log n)^{1/2})$. In addition, it is easily shown that (4.1.2) implies the following more general result describing the exact asymptotics of the probabilities of right-hand large deviations in the full range:

(4.1.3′) $$\mathbf{P}\{X_1(n) \ge a_n + b_n \cdot x\} \sim 1 - \Lambda(x) \cdot \exp\{-x^2/2a_n^2\}/(1 + x/a_n^2)$$

as $n \to \infty$, $x = x(n) \to \infty$, where $b_n := a_n^{-1}$, and a_n is the same as in (4.1.2); $a_n \sim (2 \cdot \log n)^{1/2}$ as $n \to \infty$.

Remark 4.1.1. It is obvious that in the range of deviations $x = o((\log n)^{1/2})$, the asymptotics of $\mathbf{P}\{X_1(n) > a_n + b_n \cdot x\}$ is completely determined by the first factor (compare to (4.1.3)). On the other hand, for $x \geq Const \cdot (\log n)^{1/2}$, $x = o(\log n)$, the asymptotics of $\mathbf{P}\{X_1(n) > a_n + b_n \cdot x\}$ is completely determined by $(1 - \Lambda(x)) \cdot \exp\{-x^2/2a_n^2\}$, whereas for $x \geq Const \cdot \log n$ all three factors on the right-hand side of (4.1.3′) should be considered.

In this section, we employ a different method, namely, **tail approximation**, which was briefly described in the Introduction (see formulas (0.12) - (0.12′) therein). Recall that using this method, one can construct asymptotic expansions for the probabilities of the right-hand large deviations along with accurate estimates of remainders. Note that by analogy with arguments below formula (0.8), one can drop the terms of the alternating sum on the right-hand side of (0.12) beginning with an arbitrary fixed number; the absolute value of the emerged error is known to be bounded by the absolute value of the first omitted term. In particular, dropping the terms of the alternating sum on the right-hand side of (0.12) beginning with the second one, we get that

(4.1.4) $$\mathbf{P}\{X_1(n) > y\} = n \cdot (1 - F(y)) + O((n \cdot (1 - F(y)))^2)$$

as $n \to \infty$, $y \to \infty$ such that $n \cdot (1 - F(y)) \to 0$. An application of formula (0.12) to the normal sample yields the following result:

Theorem 4.1.1. Let us assume that the i.i.d. random sequence $\{X_n, n \geq 1\}$ is the standard normal, and denote the common Laplace distribution function of X_n's by Φ. Let a_n be defined as in (4.1.2), and $b_n := a_n^{-1}$. Then for any positive x,

(4.1.5)
$$\mathbf{P}\{X_1(n) > a_n + b_n x\} = \sum_{k=1}^{n} (-1)^{k+1} \cdot \binom{n}{k} \cdot n^{-k} \cdot \left\{ e^{-x} \cdot \frac{\exp\{-x^2/2a_n^2\}}{1 + x/a_n^2} \cdot \left\{ 1 - \frac{1}{a_n^2 \cdot (1 + x/a_n^2)^2} + \ldots + (-1)^{\ell} \cdot \frac{1 \cdot 3 \cdot \ldots \cdot (2\ell - 1)}{a_n^{2\ell} \cdot (1 + x/a_n^2)^{2\ell}} + \ldots \right\} \right\}^{k}.$$

Proof of Theorem 4.1.1 is straightforward. It involves an application of (0.12) with $y = a_n + b_n \cdot x$, an expansion of $1 - \Phi(x)$ over powers of x as $x \to \infty$ (see, e.g., Feller (1971)

(vol. I, Chapter 7, Section 7, Problem 1)) and the fact that

$$\frac{n\cdot\exp\{-(a_n+b_n\cdot x)^2/2\}}{\sqrt{2\pi}(a_n+b_n\cdot x)} = e^{-x}\cdot\frac{\exp\{-x^2/2a_n^2\}}{1+x/a_n^2}. \quad \square$$

Remark 4.1.2. Note that both alternating sums on the right-hand side of (4.1.5) can be dropped at any term; the absolute value of the emerged error will be bounded by the absolute value of the first omitted term. In particular, Theorem 4.1.1 implies that

$$\begin{aligned}\mathbf{P}\{X_1(n)>a_n+b_n\cdot x\} &= \big(1-\Lambda(x)\big)\cdot\frac{\exp\{-x^2/2a_n^2\}}{1+x/a_n^2}\\ &\cdot\left\{1-\frac{1}{a_n^2\cdot(1+x/a_n^2)^2}\right\}+\mathrm{O}\left(\big(1-\Lambda(x)\big)\cdot\frac{\exp\{-x^2/2a_n^2\}}{a_n^4\cdot(1+x/a_n^2)^5}\right)\\ &+\mathrm{O}\left(\big(1-\Lambda(x)\big)^2\cdot\frac{\exp\{-x^2/2a_n^2\}}{1+x/a_n^2}\right)\end{aligned} \tag{4.1.5'}$$

as $n\to\infty$, $x\to\infty$.

Remark 4.1.3. Note that our representations (4.1.5) and (4.1.5′) are similar to Theorem 3 of Hall (1980). However, our results seem to be more convenient for computations. In addition, the proposed tail approximation method is equally applicable to an arbitrary distribution function, whereas the range of applications of Theorem 3 of Hall (1980) is confined to the normal samples. In this respect, let us quote the following remark from Cohen (1982a) (see p. 329 therein): '*Hall (1980) suggested approximations to $\Phi^n(x)$ using some refined inequalities for the normal tail function. These approximations are much closer to $\Phi^{(n)}(x)$ than the penultimate approximations. Thus if the X_i's are indeed independent and identically normally distributed and if n is known, then Hall (1980) gives better estimates of the distribution of $Y_n = \max\{X_i\}$ than approximations based on extreme value theory. However, in practice we are often uncertain of the normality, the independence and perhaps the value of n. Since the three limit laws apply to a large class of initial distributions, and often in the dependent case (cf. Galambos (1978)), extreme* *value theory approximations are more robust than the alternatives suggested by Hall (1980).'*

Now, in view of Remark 4.1.3, let us suggest the following approach, which in our opinion provides a more appropriate approximation for distributions of maxima. Hereinafter, we refer to this approach as the **tail approximation/extreme value approximation alternative**. Note that in this section, we apply this approach only to the standard normal sample, the classical test sample of the extreme value theory. Of course, the range of applications of this alternative is not confined by the normal sample.

It is natural to require the relative error $\delta(n,x)$ of the tail approximation to be less than ε (given à priori). Then by Bonferroni's inequalities (cf., e.g., Feller (1971) (Vol. I, Chapter IV, (5.7)) if

$$\mathbf{P}_n := \mathbf{P}\{X_i > a_n + b_n \cdot x\} < 1/n$$

then

$$n \cdot \mathbf{P}_n - \frac{n^2 \cdot \mathbf{P}_n^2}{2} \le \mathbf{P}\{X_1(n) > a_n + b_n \cdot x\} \le n \cdot \mathbf{P}_n.$$

Hence,

$$\delta(n,x) \le \frac{n \cdot \mathbf{P}_n}{2(1 - n \cdot \mathbf{P}_n/2)} \quad (<\epsilon).$$

Obviously, the product $n \cdot \mathbf{P}_n$ is assumed to be small enough. Making simple computations we get that for fixed ε and n, the above inequality is fulfilled (i.e. the relative error of the approximation of $\mathbf{P}\{X_1(n) > a_n + b_n \cdot x\}$ by the leftmost term on the right-hand side of (4.1.5′) is less than ε) if

$$x \ge X_1(n,\epsilon) := \frac{a_n^2}{3} \cdot \left\{ \sqrt{1 + \frac{6}{a_n^2} \cdot \log \frac{\epsilon + 1}{2\epsilon}} - 1 \right\}.$$

On the other hand, it is obvious in view of (4.1.3) and (4.1.3′) that the extreme value approximation is accurate at least for the values of x being sufficiently small compared to $(\log n)^{1/2}$. Moreover, representation (4.1.5′) implies that

$$\frac{\left|\mathbf{P}\{X_1(n) > a_n + b_n \cdot x\} - \left(1 - \Lambda(x)\right)\right|}{1 - \Lambda(x)}$$

$$\leq 1 - \frac{\exp\{-x^2/2a_n^2\}}{1 + x/a_n^2} \cdot \left\{\frac{1 - \Lambda(x)}{2} + 1\right\} \quad (< \epsilon).$$

Making simple transformations and neglecting the second order terms we get that the above inequalities are fulfilled for

$$\epsilon > x^2/2a_n^2 + x/a_n^2.$$

Hence, we obtain the result that for fixed ε and n the relative error of the extreme value approximation of $\mathbf{P}\{X_1(n) > a_n + b_n \cdot x\}$ is less than ε if

$$x \leq X_2(n,\varepsilon) := (1 + 2 \cdot a_n^2 \cdot \varepsilon)^{1/2} - 1.$$

It is not surprising that for any positive (sufficiently small) ε, and for any integer $n \geq N(\varepsilon) := [(\log (\varepsilon+1)/\varepsilon))^2 /8\varepsilon] + 1$, there is an overlap between a lower bound $X_1(n,\varepsilon)$ and an upper bound $X_2(n,\varepsilon)$. In other words, for $n \geq N(\varepsilon)$ we can always provide an approximation for the distribution of the maximum from the normal sample whose relative error is less than (given à priori) ε. Moreover, there exists a certain range of deviations in which both approximations give suitable results.

Finally, one can summarize the above discussion in the following algorithm: assume that we should find an approximation of $\mathbf{P}\{X_1(n) > a_n + b_n \cdot x\}$ for given n and x whose relative error would be less than ε. Then we can choose the tail approximation if $x \geq X_1(n,\varepsilon)$ or the extreme value approximation if $x \leq X_2(n,\varepsilon)$. This is possible at least for $n \geq N(\varepsilon)$.

Remark 4.1.4. Let us state the following open problems related to the suggested alternative.

(i) To extend the suggested alternative to a wider class of distribution (e.g., class **N** introduced in Cohen (1982b)).

(ii) To derive sharper estimates for ranges of applications of both approximations. In particular, it would be interesting to find the border function $\tilde{X}(n,\varepsilon)$ such that for $x \leq \tilde{X}(n,\varepsilon)$

the extreme value approximation should be chosen, whereas for $x \geq \tilde{X}(n,\varepsilon)$ the tail approximation should be chosen.

(iii) To derive the asymptotics of the left-hand large deviations.

Now, let us proceed with the consideration of the asymptotics of the distribution of the maximum $X_1(n)$ from the sample of i.i.d. random variables whose common distribution function $F(\cdot)$ has the right-hand tail of power type (i.e., Condition (0.1) of the Introduction is fulfilled). Recall that in this case, the results on weak convergence and on large deviations of $X_1(n)$ are given by (0.13) and (0.14), respectively. Moreover, comparing the result on weak convergence to $\Psi_\alpha(\cdot)$ with the fact that the exact asymptotics of the probabilities of large deviations of $X_1(n)$ is expressed in terms of the tail of function $\Psi_\alpha(\cdot)$, we obtain the result that the tail approximation and the extreme value approximation coincide in this case, and we do not have any alternative but the extreme value approximation.

Now, let us derive refinements to (0.13) and (0.14). Our approach resembles the approach used in the previous chapters for the derivation of refinements for $\mathbf{P}\{S_n > y\}$. Namely, if more precise information on the tail behavior of function F is available (compare to (0.1)) then more precise representations for $\mathbf{P}\{X_1(n) > y\}$ as $n \rightarrow \infty$ than (0.14) can be derived. By analogy with the results of the previous chapters, it seems reasonable to assume that the right-hand tail of F admits the following expansion over negative powers of x as $x \rightarrow \infty$ (i.e., Condition (0.5) of the Introduction is fulfilled).

On the other hand, the next result, Theorem 4.1.2, provides the asymptotic expansions for the distribution of $X_1(n)$ refining both the above mentioned result on weak convergence of F_n to Ψ_α (see (0.13)) and relationship (0.14) in the case when condition (0.5) is valid. The remainder term of the expansion of this theorem may be viewed as the unremovable error generated by the lack of the perfect information on the tail behavior of F.

Theorem 4.1.2 (cf. Vinogradov (1992b) Theorem 1). Let Condition (0.5) be fulfilled. Then

$$F_n(x) = \mathbf{P}\{X_1(n) \le (c_{\alpha_1} \cdot n)^{1/\alpha_1} \cdot x\} = \left(1 - n^{-1} \cdot x^{-\alpha_1}\right)^n$$

$$\cdot \left(1 + \sum_{m=1}^{[r/\alpha_2] \vee [(r-\alpha_1)/(\alpha_2-\alpha_1)]} \frac{(-1)^m}{m!} \cdot \left(n \cdot \sum_{s=1}^{[r/\alpha_2]} \frac{1}{s} \left(\sum_{i=2}^{\ell} \frac{c_{\alpha_i}}{c_{\alpha_1}^{\alpha_i/\alpha_1}} \cdot n^{\alpha_i/\alpha_1} x^{-\alpha_i}\right)^s \right.\right.$$

$$\left.\left. \cdot \sum_{k=0}^{[(r/m - s\cdot\alpha_2)/\alpha_1] \vee 0} (-1)^k \binom{-s}{k} n^{-k} x^{-k\alpha_1} \right)^m \right) \cdot \left(1 + \delta(x,n) \cdot n^{-(r-\alpha_1)/\alpha_1} \cdot x^{-r/\alpha_1}\right),$$

where $\delta(x,n) \to 0$ as $n \to \infty$ uniformly on the rays $[C, +\infty)$; $\binom{-s}{0} := 1$, and

$$\binom{-s}{k} := (-1)^k \cdot s \cdot (s+1) \cdot \ldots \cdot (s+k-1)/k!$$

for integer $k \ge 1$ (Here, $C > 0$ is fixed.)

Proof of Theorem 4.1.2 essentially consists of an application of (0.12) and (0.5). □

Let us emphasize that the comparison of the above results related to maxima from normal samples with those related to maxima from samples of distributions having right-hand tails of the power type reveals the presence of two polar types of the limiting behavior of the probabilities of large deviations.

Example 4.1.1 (compare to Example 1 in Section 6 of Smith (1982) and Exercise 2.4.2 of Resnick (1987)). Let Condition (0.5) be fulfilled with $\ell = 2$, $\alpha_2 < 2\cdot\alpha_1$, and $r = \alpha_2$. Then

$$F_n(x) = \left(1 - \frac{x^{-\alpha_1}}{n}\right)^n \cdot \left(1 - \frac{c_{\alpha_2}}{c_{\alpha_1}^{\alpha_2/\alpha_1}} \cdot n^{-(\alpha_2-\alpha_1)/\alpha_1} \cdot x^{-\alpha_2}\right)$$

$$\cdot \left(1 + \delta(x,n) \cdot n^{-(\alpha_2-\alpha_1)/\alpha_1} \cdot x^{-\alpha_2}\right),$$

where $\delta(x,n) \to 0$ as $n \to \infty$ uniformly on the rays $[C,+\infty)$ with $C > 0$. Note that in contrast to Smith (1982), our Theorem 4.1.2 provides a non-uniform estimate of the remainder and involves a more accurate principal term $(1 - n^{-1}x^{-\alpha_1})^n$ instead of the approximation in terms of the limiting distribution $\Psi_{\alpha_1}(x)$ used by Smith.

Now, let us derive the asymptotics of the probabilities of large deviations for the maximum $X_1(n)$. Recall relationship (4.1.4) that is valid for an arbitrary random sequence $\{X_n, n \geq 1\}$ and apply it for the case of fulfilment of Condition (0.5). Then one gets the following:

Example 4.1.2. Let Condition (0.5) be fulfilled. Then it follows from (4.1.4) that

$$\mathbf{P}\{X_1(n) > (c_{\alpha_1} \cdot n)^{1/\alpha_1} \cdot x\}$$

(4.1.6)

$$= \sum_{\substack{1 \leq i \leq l \\ \alpha_i < 2\alpha_1}} \frac{c_{\alpha_i}}{c^{\alpha_i/\alpha_1}} \cdot n^{-(\alpha_i - \alpha_1)/\alpha_1} \cdot x^{-\alpha_i} + O\left(x^{-2\cdot\alpha_1}\right)$$

as $n \to \infty$, $x \to \infty$.

Note that the comparison of the remainder of the expansion of Theorem 4.1.2 with the remainder presented on the right-hand side of (4.1.6) implies that in the considered special case, the expansion of Theorem 4.1.2 is more precise at least in the range of deviations in which x grows to infinity slower than any power of n.

It should be mentioned that one can construct the asymptotic expansion in the theorem on weak convergence of the properly normalized distribution of the minimum $X_n(n)$ to the Weibull distribution $W_\alpha(\cdot)$ by employing the same technique (Recall that the Weibull distribution $W_\alpha(\cdot)$ is defined as $W_\alpha(x) := 1 - \exp\{-x^\alpha\}$ if $x > 0$; $W_\alpha(x) := 0$ otherwise.) We also need to assume that the common distribution function $F(\cdot)$ of the random sample $\{X_n, n \geq 1\}$ satisfies the following condition:

$$(4.1.7) \qquad F(x) = \sum_{i=1}^{l} c_{\alpha_i} \cdot x^{\alpha_i} + o(x^r)$$

as $x \to 0$, where $c_\alpha > 0$, and $0 < \alpha_1 < \alpha_2 < \ldots < \alpha_l \leq r$.

Let us now formulate the corresponding rigorous result skipping the proof (the proof is in fact very simple and relies on the transformation $x \to 1/x$ that exchanges order statistics and converts Condition (4.1.7) to Condition (0.5) with a subsequent application of Theorem 4.1.2).

Theorem 4.1.3. Let Condition (4.1.7) be fulfilled. Then

$$\mathbf{P}\{(c_{\alpha_1}\cdot n)^{1/\alpha_1}\cdot X_n(n)\le x\} = 1-\left(1-n^{-1}\cdot x^{\alpha_1}\right)^n$$

$$\cdot\left(1+\sum_{m=1}^{[r/\alpha_2]\vee[(r-\alpha_1)/(\alpha_2-\alpha_1)]}\frac{(-1)^m}{m!}\cdot\left(n\cdot\sum_{s=1}^{[r/\alpha_2]}\frac{1}{s}\left(\sum_{i=2}^{\ell}\frac{c_{\alpha_i}}{c_{\alpha_1}^{\alpha_i/\alpha_1}}\cdot n^{-\alpha_i/\alpha_1}x^{\alpha_i}\right)^s\right.\right.$$

$$\left.\left.\cdot\sum_{k=0}^{[(r/m-s\cdot\alpha_2)/\alpha_1]\vee 0}(-1)^k\binom{-s}{k}n^{-k}x^{k\alpha_1}\right)^m\right)\cdot\left(1+\delta(x,n)\cdot n^{-(r-\alpha_1)/\alpha_1}\cdot x^{r/\alpha_1}\right),$$

where $\delta(x,n)\to 0$ as $n\to\infty$ uniformly on the intervals $[0,C\,]$.

Remark 4.1.5. It is clear that an application of Theorem 4.1.3 enables one to derive the asymptotic expansion for distribution of the maximum $X_1(n)$ in the case, where

$$1 - F(x) = \sum_{i=1}^{\ell} c_{\alpha_i}\cdot(a-x)^{\alpha_i} + o\big((a-x)^r\big)$$

as $x \uparrow a$, where $a<\infty$, $c_{\alpha_i}>0$, and $0<\alpha_1<\alpha_2<\ldots<\alpha_\ell\le r$. Note that distributions of such type often arise in various statistical estimation problems (cf., e.g., Ibragimov and Has'minskii (1981) for details).

4.2 Large Deviations for Trimmed Sums.

In this section, we state and prove Theorems 4.2.1 - 4.2.5. Note that since the proofs of all these theorems follow along the same lines we only give the complete proof of Theorem 4.2.1 and describe modifications that should be made in the proofs of the other theorems.

First we give some auxiliary notation. Let us say that a function $f\in \mathbf{D}[0,1]$ makes a *jump* at a point $t\in(0,1]$ if $f(t)\neq f(t-)$. Let several time instants $0<t_1<\ldots<t_k\le 1$ and real numbers $x_1, \ldots, x_k$ such that $x_1\cdot\ldots\cdot x_k\neq 0$ be fixed. Following Wentzell (1985) we denote a piecewise constant function (defined on the unit interval) equal to zero as $s\in[0,t_1)$, to x_1 as $s\in[t_1,t_2),\ldots$, to x_k as $s\in[t_k,1]$ by

$$0^{t_1}x_1^{t_2}\ldots x_{k-1}^{t_k}x_k.$$

Let us denote the set of all such functions by $\mathbf{B}^k$, where $k\ge 0$ is any fixed integer. In particular, $\mathbf{B}^0$ consists of only one function equal to zero identically. Let ∂A and $(A)_\varepsilon$ denote the boundary of the set A and its ε-neighborhood, respectively.

In this section, we will use the concepts of **w-continuous** and **w-saturated** sets (which are due to Pinelis (1981)) in the context of the standard univariate Wiener process $\mathbf{w}(\cdot)$.

Definition 4.2.1. A Borel set G is called **w-continuous** if for any real $c > 0$,

$$\mathbf{P}\{\mathbf{w}(\cdot) \in u \cdot (\partial G)_\tau\} = o(\mathbf{P}\{\mathbf{w}(\cdot) \in G\})$$

as $u \to \infty$ with $\tau = \exp\{-c \cdot u^2\}$.

Definition 4.2.2. A Borel set $\aleph$ is called **w-saturated** if

$$\log \mathbf{P}\{\mathbf{w}(\cdot) \in u \cdot G\} = O(u^2)$$

as $u \to \infty$. Now, let us introduce the following set from $\mathbf{D}[0,1]$:

$$G_{a,b} := \{f \in \mathbf{D}[0,1] : f(1) - \sup_{t \in [0,1]} \big(f(t) - f(t-)\big) > a, f(1) > b\}, \tag{4.2.1}$$

where $0 < a < b$. It is clear that

$$\mathbf{P}\{S_n - X_1(n) > ay, S_n > by\} = \mathbf{P}\{S_{n,y} \in G_{a,b}\}. \tag{4.2.2}$$

Let

$$r := \rho(G_{a,b}, \mathbf{B}^1),$$

where ρ stands for the uniform metric on $\mathbf{D}[0,1]$. It can be easily shown that $r > 0$. Let us also note that the sets $G_{a,b}$ are w-continuous and w-saturated.

The first result of this section generalizes Theorem 1 of Vinogradov and Godovan'chuk (1989), where a much stronger condition (0.1′) was assumed instead of Condition (0.10) that is posed below. In the following Theorem 4.2.1, we describe the exact asymptotics of the probability of r.s.f. $S_{n,y}$ to hit the set $G_{a,b}$, in the full range of large deviations.

Theorem 4.2.1. Let Conditions (0.1) and (0.10) be fulfilled with $\alpha_1 = \beta > 2$, $\mathbf{E}X_i = 0$, and $\mathbf{V}X_i = 1$. Let $0 < a < b < \infty$. Then

$$\begin{aligned} \mathbf{P}\{S_{n,y} \in G_{a,b}\} &= \mathbf{P}\{\sqrt{n} \cdot w(1) > by\} \cdot (1 + o(1)) \\ &+ n \cdot c_{\alpha_1} \cdot (b-a)^{-\alpha_1} \cdot y^{-\alpha_1} \cdot \mathbf{P}\{\sqrt{n-1} \cdot w(1) > ay\} \cdot (1 + o(1)) \\ &+ \binom{n}{2} \cdot c_{\alpha_1}^2 \cdot y^{-2\alpha_1} \cdot \big(\mu(a,b) + o(1)\big) \end{aligned} \tag{4.2.3}$$

as $n \to \infty$, $y/\sqrt{n} \to \infty$, where $\mu(a,b)$ is defined as follows:

$$(4.2.4) \qquad \mu(a,b) = \begin{cases} a^{-2\alpha_1} & \text{if } b \le 2a; \\ a^{-\alpha_1}\cdot(b-a)^{-\alpha_1} + \int_a^{b-a} (b-x)^{-\alpha_1}\cdot d(-x^{-\alpha_1}) & \text{otherwise.} \end{cases}$$

Remark 4.2.1. For each summand on the right-hand side of (4.2.3), there exists its own range of deviations in which this summand makes the main contribution to the sum (i.e., the other two terms are negligible). To explain this, we consider a special case $a = 1$ and $b = 3$. We have that for any real $\kappa > 0$, if

$$y \le \sqrt{\frac{(\alpha_1 - 2 - \kappa)\cdot n \cdot \log n}{8}}$$

then the first term makes the main contribution to the sum on the right-hand side of (4.2.3), and the event $\{S_{n,y} \in G_{a,b}\}$ occurs mostly due to close-to-continuous paths. The asymptotics of $\mathbf{P}\{S_{n,y} \in G_{a,b}\}$ up to logarithmic equivalence in this range of deviations can be expressed in terms of the action functional (cf., e.g., Wentzell (1990) Theorem 4.4.3 and also Section 5.3 of the present monograph). Typical paths of r.s.f. $S_{n,y}$ are shown in Fig. 1.

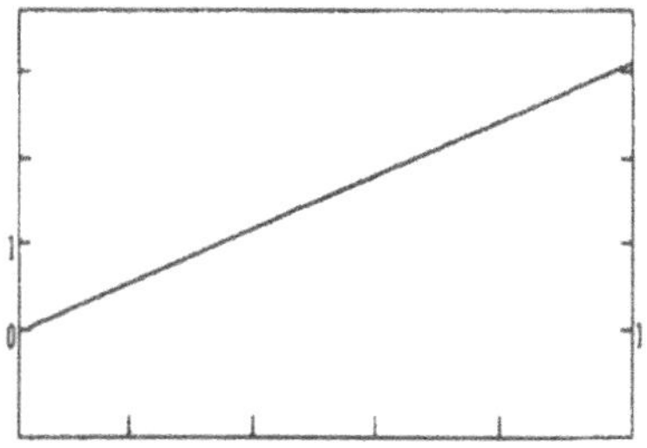

Figure 1

On the other hand, if

$$\sqrt{\frac{(\alpha_1 - 2 + \kappa)\cdot n \cdot \log n}{8}} \le y \le \sqrt{(\alpha_1 - 2 - \kappa)\cdot n \cdot \log n}$$

then the second term is the greatest, and the typical paths are shown in Fig. 2. Namely, r.s.f. $S_{n,y}$ makes one big jump (at a random time instant uniformly distributed over the unit interval) with magnitude greater than $(b - a)$. In addition, if we glue the typical path of $S_{n,y}$ at the point of the maximal jump, then the modified trajectory of r.s.f. $S_{n,y}$

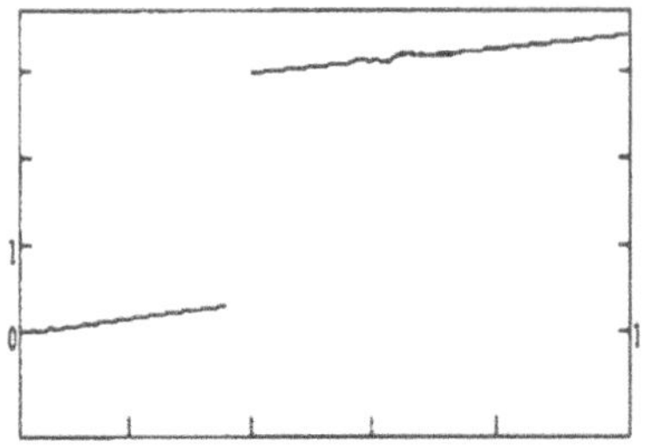

Figure 2

would resemble that of the Wiener process $\mathbf{w}(\cdot)$. Finally, if

$$y \geq \sqrt{(\alpha_1 - 2 + \kappa)\cdot n \cdot \log n}$$

then the main contribution to the sum comes from the third term, and the paths which determine the asymptotics of the probability of event $\{S_{n,y} \in G_{a,b}\}$ are piecewise constant and are presented in Fig. 3.

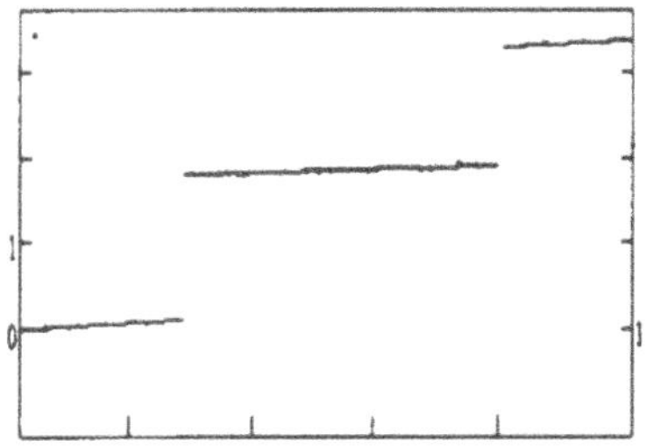

Figure 3

Proof of Theorem 4.2.1. Let $\beta = \beta(n)$ be a bounded sequence of positive real numbers such that $\beta(n)\cdot n^{\delta} \to \infty$ as $n \to \infty$ for each $\delta > 0$. In the sequel, we will choose a specific sequence $\beta(n)$ satisfying these properties (see Lemma 4.2.4 below).

Let us take an arbitrary $\gamma \in (0,(a/8)\wedge(r/40)\wedge(b-a)/2)$ and introduce the following quantities:

$$\Pi_1 := \mathbf{P}\{S_{n,y} \in G_{a,b};\ |X_i| \leq \beta n/y,\ i = 1, \dots, n\};$$

$$\Pi_2 = \mathbf{P}\Big(\bigcup_{i=1}^{n}\{|X_i| > \beta n/y\}\Big);$$

$$\Pi_3 = \sum_{i=1}^{n} \mathbf{P}\{S_{n,y} \in G_{a,b};\, |X_i| > \beta n/y,\ |X_j| \le \beta n/y\ \forall j \ne i\};$$

$$\Pi_4 := \sum_{1\le i\ne j\le n} \mathbf{P}\{S_{n,y} \in G_{a,b};\, |X_i| > \beta n/y,\ |X_j| > \beta n/y,\ |X_k| \le \gamma y \forall k \ne i\};$$

$$\Pi_5 := \sum_{1\le i<j\le n} \mathbf{P}\{S_{n,y} \in G_{a,b};\, |X_i| > \gamma y,\ |X_j| > \gamma y\};$$

$$\Pi_6 := \sum_{i=1}^{n} \mathbf{P}\{S_{n,y} \in G_{a,b};\, |X_i| > \gamma y,\ |X_j| \le \gamma y\ \forall j \ne i\};$$

$$\Pi_7 := \mathbf{P}\Big(\bigcup_{1\le i<j<k\le n} \{|X_i| \wedge |X_j| \wedge |X_k| > \gamma y\}\Big);$$

$$\Pi_8 := \mathbf{P}\{S_{n,y} \in G_{a,b};\, |X_i| \le \gamma y,\ i = 1,\dots,n\}.$$

Now, we proceed with a series of lemmas whose combinations will imply the assertion of Theorem 4.2.1.

Lemma 4.2.1 (cf. Pinelis (1981) Lemma 9). For any positive M, there exists a positive constant c such that

$$\Pi_2 = O(\exp\{-M\cdot y^2/n\})$$

as $n \to \infty$ with $n^{1/2} \le y \le (c\cdot n\cdot \log n)^{1/2}$.

Lemma 4.2.2 (cf. Pinelis (1981) Lemma 4). There exist positive constants c_0 and δ such that for any subset U of $\mathbf{D}[0,1]$ (for which the probabilities of the events $\{S_{n,y}(\cdot) \in U\}$ and $\{n^{1/2}\, w(\cdot) \in y\cdot U\}$ are well-defined), the following representation is valid:

$$\mathbf{P}\{S_{n,y}(\cdot) \in U,\ |X_i| \le \beta n/y,\ i = 1,\dots,n\} = \mathbf{P}\{\sqrt{n}\cdot w(\cdot) \in y\cdot U\}\cdot(1 + O(n^{-\delta})$$

$$+\ O\big(\mathbf{P}\{\sqrt{n}\cdot w(\cdot) \in y\cdot(\partial U)_\beta\}\big) + O(\exp\{-c_0\cdot y^2/n\})$$

as $n \to \infty$, $y/\sqrt{n} \to \infty$ with $y \le n^{1/2+\delta}$.

Lemma 4.2.3 (compare to Pinelis (1981) remark to formula (1)). Fix arbitrary $0 < a_1 < a_2 < \infty$. Then

$$\mathbf{P}\{u\cdot(a-\eta) \le w(1) \le u\cdot(a+\eta)\} = o(\mathbf{P}\{w(1) \ge u\cdot a\})$$

uniformly in $a \in [a_1, a_2]$, as $u \to \infty$ with $\eta = o(u^{-2})$.

The following lemma plays the key role in the proof of Theorem 4.2.1. Owing to its statement the asymptotics of the probabilities of large deviations acquires the term of a new type (see also Remark 4.2.1 to Theorem 4.2.1).

Lemma 4.2.4. Let $\beta(n)$ be a bounded positive sequence such that $\beta(n) := o(1/\log n)$ as $n \to \infty$. Choose arbitrary $0 < c < C < \infty$. Then

$$\Pi_3 \sim n \cdot c_{\alpha_1} \cdot (b-a)^{-\alpha_1} \cdot y^{-\alpha_1} \cdot \mathbf{P}\{\sqrt{n-1} \cdot w(1) > ay\}$$

as $n \to \infty$ with $(cn \cdot \log n)^{1/2} \leq y \leq (Cn \cdot \log n)^{1/2}$.

Proof of Lemma 4.2.4. It follows from the definition of Π_3 that

$$\Pi_3 = \sum_{i=1}^{n} \mathbf{P}\{S_n - X_1(n) > ay, S_n > by, X_i > \beta n/y, |X_j| \leq \beta n/y \quad \forall j \neq i\}$$

$$+ \sum_{i=1}^{n} \mathbf{P}\{S_n - X_1(n) > ay, S_n > by, X_i < -\beta n/y, |X_j| \leq \beta n/y \quad \forall j \neq i\}$$

$$= \sum_{i=1}^{n} \mathbf{P}\{S_n - X_i > ay, S_n > by, X_i > \beta n/y, |X_j| \leq \beta n/y \quad \forall j \neq i\}$$

$$+ \sum_{i=1}^{n} \mathbf{P}\{S_n - X_1(n) - X_i > ay - X_i, S_n - X_i > by - X_i, \quad X_i < -\beta n/y,$$

$$|X_j| \leq \beta n/y \quad \forall j \neq i\}.$$

Therefore,

$$\Pi_3 = n \cdot \int_{\beta n/y}^{\infty} \mathbf{P}\{S_{n-1} > ay, S_{n-1} > by - z, |X_j| \leq \beta n/y, i = 1, \ldots, n-1\} \cdot dF(z)$$

$$+ n \cdot \int_{-\infty}^{-\beta n/y} \mathbf{P}\{S_{n-1} - X_1(n) > ay - z, S_{n-1} > by - z, |X_j| \leq \beta n/y, i = 1, \ldots, n-1\} \cdot dF(z).$$

Splitting the first integral on the right-hand side of the above representation for Π_3 into the sum of two others, one gets that this integral is equal to

$$\int_{\beta n/y}^{(b-a)y} \mathbf{P}\{S_{n-1} > by - z, |X_j| \leq \beta n/y, i = 1, \ldots, n-1\} \cdot dF(z)$$

$$
\begin{aligned}
& + \int_{(b-a)y}^{\infty} \mathbf{P}\{S_{n-1} > ay,\ |X_j| \le \beta n/y,\ i=1,\dots,n-1\} \cdot dF(z) \\
(4.2.5) \qquad & = \int_{\beta n/y}^{(b-a)y} \mathbf{P}\{S_{n-1} > by - z,\ |X_j| \le \beta n/y,\ i=1,\dots,n-1\} \cdot dF(z) \\
& + \mathbf{P}\{S_{n-1} > ay,\ |X_j| \le \beta n/y,\ i=1,\dots,n-1\} \cdot \mathbf{P}\{X_n > (b-a)y\}.
\end{aligned}
$$

A subsequent application of Lemmas 4.2.2 - 4.2.3 implies that

$$
\begin{aligned}
& \mathbf{P}\{S_{n-1} > ay,\ |X_j| \le \beta n/y,\ i=1,\dots,n-1\} \cdot \mathbf{P}\{X_n > (b-a)y\} \\
(4.2.6) \qquad & \sim \mathbf{P}\{X_1 > (b-a)y\} \cdot \mathbf{P}\{\sqrt{n-1} \cdot w(1) > ay\}.
\end{aligned}
$$

Now, estimating the integral from $\beta n/y$ to $(b-a)y$ on the right-hand side of (4.2.5) by the use of Lemma 4.2.2 we obtain that this integral is equal to

$$
\begin{aligned}
& \int_{\beta n/y}^{(b-a)y} \Big\{ \mathbf{P}\{\sqrt{n-1} \cdot w(1) > by - z\} \cdot \big(1 + O(n^{-\delta})\big) + O\big(\mathbf{P}\{y \cdot (b - z/y - \beta) \\
(4.2.7) \qquad & \le \sqrt{n-1} \cdot w(1) \le y \cdot (b - z/y + \beta)\big) + O\big(\exp\{-c_0 \cdot y^2/n\}\big) \Big\} \cdot dF(z).
\end{aligned}
$$

To estimate the second term in the integrand of the integral (4.2.7), we apply Lemma 4.2.3. One gets that this term is

$$
o\big(\mathbf{P}\{\sqrt{n-1} \cdot w(1) > by - z\}\big)
$$

uniformly within the integration range $[\beta n/y,\ (b-a)y]$. Hence, the integral (4.2.7) can be rewritten as

$$
(4.2.8) \qquad \int_{\beta n/y}^{(b-a)y} \mathbf{P}\{\sqrt{n-1} \cdot w(1) > by - z\} \cdot \big(1 + o(1)\big) \cdot dF(z).
$$

Now, let us demonstrate that the integral (4.2.8) is negligible compare to the expression on the right-hand side of (4.2.6). Indeed, on can use the well-known upper estimate for the tail of the standard normal distribution to get that for sufficiently large n, the integral

(4.2.8) does not exceed

$$(4.2.9) \qquad \frac{2}{\sqrt{2\pi}} \cdot \int_{\beta n/y}^{(b-a)y} \frac{\sqrt{n}}{by-z} \cdot \exp\left\{-\frac{by-z}{2(n-1)}\right\} \cdot dF(z).$$

Fix an arbitrary δ such that $0 < \delta < (b-a)$ and split the integral presented in (4.2.9) into the sum of integrals from $\beta n/y$ to $(b-a-\delta)y$ and from $(b-a-\delta)y$ to $(b-a)y$. It is clear that the integral from $\beta n/y$ to $(b-a-\delta)y$ does not exceed

$$\frac{\exp\left\{-(a+\delta)^2 y^2/\left(2(n-1)\right)\right\}}{(a+\delta)y/\sqrt{n-1}} \cdot \left(1 - F(\beta n/y)\right).$$

In addition, it is relatively easy to see that there exists a positive number μ belonging to the integration range from $\beta n/y$ to $(b-a-\delta)y$ and depending on δ such that

$$(4.2.10) \qquad \begin{aligned} \frac{\exp\left\{-(a+\delta)^2 y^2/\left(2(n-1)\right)\right\}}{(a+\delta)y/\sqrt{n-1}} &= o\left(n^{-\mu} \cdot \frac{\exp\left\{-a^2 y^2/\left(2(n-1)\right)\right\}}{y/\sqrt{n-1}}\right) \\ &= o\left(n^{-\mu} \cdot \mathbf{P}\{\sqrt{n-1} \cdot w(1) > ay\}\right). \end{aligned}$$

Now, recall that $\beta(n) := o(1/\log n)$ and that the asymptotics of the right-hand tail of $F(x)$ is given by (0.1). Therefore,

$$(4.2.11) \qquad \left(1 - F(\beta n/y)\right) = o\left(n^{-\mu} \cdot \left(1 - F\left((b-a)y\right)\right)\right).$$

A combination of (4.2.10) and (4.2.11) implies that the integral from $\beta n/y$ to $(b-a-\delta)y$ is negligible compare to the expression on the right-hand side of (4.2.6).

It is clear that the integral from $(b-a-\delta)y$ to $(b-a)y$ does not exceed

$$(4.2.12) \qquad 4 \cdot \mathbf{P}\{\sqrt{n-1} \cdot w(1) > ay\} \cdot \left(F\left((a+\delta)y\right) - F(ay)\right).$$

In turn, expression (4.2.12) can be made negligible compare to the expression on the right-hand side of (4.2.6) on the reason that due to (0.1), we have that for any real $\varepsilon > 0$ there exist $\delta = \delta(\varepsilon) > 0$ and $Y < \infty$ such that

$$(F((a+\delta)y) - F(ay))/(1 - F((b-a)y)) < \varepsilon.$$

Hence, given $\varepsilon > 0$ we set $\delta = \delta(\varepsilon)$ in all the arguments of the lemma. A combination of the latter bound with (4.2.12) implies that the integral from $(b-a-\delta)y$ to $(b-a)y$ is also negligible compare to the expression on the right-hand side of (4.2.6). Therefore, both expression (4.2.9) and integral (4.2.8) are negligible compare to the expression on the right-hand side of (4.2.6). Hence,

$$n\cdot\int_{\beta n/y}^{\infty}\mathbf{P}\{S_{n-1}>ay,S_{n-1}>by-z,|X_j|\le\beta n/y,i=1,\ldots,n-1\}\cdot dF(z)$$

(4.2.13)

$$\sim n\cdot\mathbf{P}\{X_1>(b-a)y\}\cdot\mathbf{P}\{\sqrt{n-1}\cdot w(1)>ay\}.$$

Now, let us estimate the second term

$$n\cdot\int_{-\infty}^{-\beta n/y}\mathbf{P}\{S_{n-1}-X_1(n)>ay-z,S_{n-1}>by-z,|X_j|\le\beta n/y,i=1,\ldots,n-1\}\cdot dF(z)$$

on the right-hand side of the representation for Π_3. Getting rid of the condition $\{S_{n-1}-X_1(n)>ay-z\}$ and keeping in mind that z takes on negative values we conclude that the above expression does not exceed

$$n\cdot\int_{-\infty}^{-\beta n/y}\mathbf{P}\{S_{n-1}>by,|X_j|\le\beta n/y,i=1,\ldots,n-1\}\cdot dF(z).$$

This expression in turn is

$$\mathrm{O}\left(n\cdot\mathbf{P}\{X_1<-\beta n/y\}\cdot\mathbf{P}\{\sqrt{n-1}\cdot w(1)>by\}\right)$$

by Lemmas 4.2.2 and 4.2.3.

The proof of the fact that the latter expression is negligible compared with

$$n\cdot\mathbf{P}\{X_1>(b-a)y\}\cdot\mathbf{P}\{(n-1)^{1/2}\cdot w(1)>ay\}$$

is straightforward and essentially analogous to the derivation of estimates (4.2.10) - (4.2.11), which implies the assertion of the lemma. □

The following four technical lemmas enable one to neglect the quantities Π_4, Π_6, Π_7, and Π_8 (compare to Π_1, Π_3, and Π_5) in the proof of Theorem 4.2.1.

Lemma 4.2.5. There exists a positive constant C such that

$$\text{II}_8 = o\left(n^2 \cdot y^{-2\alpha_1}\right)$$

as n → ∞ with $y \ge (Cn \cdot \log n)^{1/2}$.

Proof of Lemma 4.2.5. Firstly, one easily obtains that

$$\text{II}_8 = \mathbf{P}\{S_n - X_1(n) > ay, S_n > by, |X_i| \le \gamma y, i=1,\dots,n\}$$

(4.2.14)

$$\le \mathbf{P}\left\{\max_{1 \le i \le n} |S_i| > by, |X_i| \le \gamma y, i=1,\dots,n\right\}$$

by getting rid of the condition $\{S_n - X_1(n) > ay\}$. The latter probability can be estimated by analogy with Lemma 1.1.1 of Section 1.1. Namely, for any integer $n \ge 1$ and $\ell \ge 1$ and for any real $y > 0$,

$$\mathbf{P}\left\{\max_{1 \le i \le n} |S_i| > by, |X_i| \le \gamma y, i=1,\dots,n\right\}$$

(4.2.15)

$$\le \mathbf{P}\left\{\max_{1 \le i \le n} |S_i| > by/(2\ell-1)\right\}^{\ell}.$$

Note that the only difference is that r.v.'s S_i and X_i in formula (1.1.3) are replaced by their absolute values in (4.2.15). Recall that $0 < \gamma < a/(3\ell-1) < b/(3\ell-1)$. A subsequent application of Lemma 1.1.4 of Section 1.1 implies that for any integer $n \ge 1$ and for any real $y > 0$, the probability on the right-hand side of (4.2.15) does not exceed

$$2^{\ell} \cdot \max_{1 \le i \le n} \mathbf{P}\{|S_i| > by/(3\ell-1) - \sqrt{2n}\}^{\ell}$$

(4.2.16)

$$\le 2^{\ell} \cdot \max_{1 \le i \le n} \mathbf{P}\left\{|S_i| > by/\left(2 \cdot (3\ell-1)\right)\right\}^{\ell}$$

for $y \ge 2 \cdot (2n)^{1/2} \cdot (3\ell-1)/b$.

Now, set $\ell = 3$ and combine (4.2.14) - (4.2.16). We get that

(4.2.17) $$\text{II}_8 \le 8 \cdot \max_{1 \le i \le n} \mathbf{P}\{|S_i| > by/16\}^3.$$

To estimate the probability on the right-hand side of (4.2.17), we apply proposition 1.1.2 with $\beta = \alpha_1$. One easily obtains that there exist positive constants C_1 and C_2 such that for any integer $n \ge 1$, for any integer $1 \le i \le n$, and for $y \ge (K_2 n \cdot \log n)^{1/2}$,

(4.2.18) $\mathbf{P}\{|S_i| > by/16\} \leq C_1 \cdot n \cdot y^{-\alpha_1}.$

The assertion of Lemma 4.2.5 then follows from (4.2.17) and (4.2.18). □

Lemma 4.2.6. For any integer $n \geq 1$ and for any real $y > 0$,

$$\Pi_7 = O\left(n^3 \cdot y^{-3\alpha_1}\right).$$

Proof of Lemma 4.2.6 is trivial. □

Lemma 4.2.7. There exists a positive constant C such that

$$\Pi_6 = o\left(n^2 \cdot y^{-2\alpha_1}\right)$$

as $n \to \infty$ with $y \geq (C n \cdot \log n)^{1/2}$.

Proof of Lemma 4.2.7. Recall that

$$r = \rho(G_{a,b}, \mathbf{B}^1) > 0.$$

For this reason, for any $i = 1, \ldots, n$, the $(r/2)$-neighborhood of the function $0^{i/n} X_i/y$ does not intersect with the set $G_{a,b}$. Hence, if the event

$$\{S_{n,y} \in G_{a,b};\ |X_i| > \gamma y,\ |X_j| \leq \gamma y \ \ \forall\, j \neq i\}$$

occurred, then either

$$\max_{1 \leq k < i} \Big| \sum_{j=1}^{k} X_j/y \Big| > r/4$$

or

$$\max_{i < k \leq n} \Big| \sum_{j=i+1}^{k} X_j/y \Big| > r/4.$$

Therefore,

$$\Pi_6 \leq 2n \cdot \mathbf{P}\{|X_n| > \gamma y\} \cdot \mathbf{P}\{\max_{1 \leq i \leq n-1} |S_i| > ry/4;\ |X_j| \leq \gamma y, j = 1, \ldots, n-1\}.$$

Note that we have already estimated the latter factor in Lemma 4.2.5 (see formulas (4.2.15) - (4.2.18) above), which implies the required estimate. □

Lemma 4.2.8. Choose arbitrary $0 < c < C < \infty$. Then

$$\Pi_4 = o\left(n^2 \cdot y^{-2\alpha_1}\right)$$

as $n \to \infty$ with $(c n \cdot \log n)^{1/2} \leq y \leq (C n \cdot \log n)^{1/3}$.

Proof of Lemma 4.2.8 is analogous to that of Lemma 4.2.7. □

The following lemma is of particular value.

Lemma 4.2.9. There exists a positive constant C such that

$$\Pi_5 \sim \binom{n}{2} \cdot c_{\alpha_1}^2 \cdot y^{-2\alpha_1} \cdot \mu(a,b)$$

as $n \to \infty$ with $y \geq (Cn \cdot \log n)^{1/2}$.

Proof of Lemma 4.2.9. Obviously,

$$\Pi_5 = \sum_{1 \leq i < j \leq n}^{n} \mathbf{P}\{S_n - X_1(n) > ay, S_n > by, X_i \wedge X_j > \gamma y\}$$

(4.2.19)

$$+ \sum_{1 \leq i < j \leq n}^{n} \mathbf{P}\{S_n - X_1(n) > ay, S_n > by, |X_i| \wedge |X_j| > \gamma y, X_i \wedge X_j < -\gamma y\}.$$

It is easily seen that the first sum on the right-hand side of (4.2.19) is equal to

$$\binom{n}{2} \cdot \mathbf{P}\{S_{n-2} + X_{n-1} \wedge X_n > ay, S_n > by, X_{n-1} \wedge X_n > \gamma y\}$$

(4.2.20)

$$+ \sum_{1 \leq i < j \leq n}^{n} \mathbf{P}\{S_n - X_1(n) > ay, S_n > by, X_i \wedge X_j > \gamma y, i \neq i_1 \neq j\}.$$

Clearly, the rightmost sum in (4.2.20) does not exceed

(4.2.21) $$\sum_{1 \leq i < j \leq n}^{n} \mathbf{P}\{X_i \wedge X_j \wedge X_k > \gamma y\} = O\left(n^3 \cdot y^{-3\alpha_1}\right).$$

On the other hand, the leftmost summand in (4.2.20) is equal to

$$\binom{n}{2} \cdot \Bigg(\int_{-\infty}^{(a-\gamma)y} \mathbf{P}\{X_{n-1} \wedge X_n > ay - z, X_{n-1} + X_n > by - z\} \cdot dF_{S_{n-2}}(z)$$

$$+ \int_{(a-\gamma)y}^{(b-2\gamma)y} \mathbf{P}\{X_{n-1} \wedge X_n > \gamma y, X_{n-1} + X_n > by - z\} \cdot dF_{S_{n-2}}(z)$$

$$+ \mathbf{P}\{X_{n-1} \wedge X_n > \gamma y\} \cdot \mathbf{P}\{S_{n-2} > (b-2\gamma)y\}\Bigg).$$

As the sum of the second and the third terms of the expression within the braces does not exceed

$$\mathbf{P}\{X_{n-1}\wedge X_n > \gamma y\}\cdot\mathbf{P}\{S_{n-2} > (a-\gamma)y\} = O\left(n\cdot y^{-3\alpha_1}\right),$$

so the first sum on the right-hand side of (4.2.19) is equal to

$$\binom{n}{2}\cdot\Bigg(\int_{-\infty}^{(a-\gamma)y}\mathbf{P}\{X_{n-1}\wedge X_n > ay, X_{n-1}+X_n > by\}\cdot dF_{S_{n-2}}(z)$$

$$+\int_{-\infty}^{(a-\gamma)y}\Big(\mathbf{P}\{X_{n-1}\wedge X_n > ay-z, X_{n-1}+X_n > by-z\}$$

$$-\mathbf{P}\{X_{n-1}\wedge X_n > ay, X_{n-1}+X_n > by\}\Big)\cdot dF_{S_{n-2}}(z) + o\left(n^2\cdot y^{-2\alpha_1}\right).$$

The absolute value of the second integral within the braces is

$$o\left(y^{-2\alpha_1}\right)$$

as $y \to \infty$. The proof of this fact is straightforward and is essentially based on the use of (0.1). Similar computations can be found in Section 1.1 (see formulas (1.1.12) - (1.1.14) therein). On the other hand, it is relatively easy to see that the first integral within the braces is equal to

$$\mathbf{P}\{X_{n-1}\wedge X_n > ay, X_{n-1}+X_n > by\}\cdot\left(1-\mathbf{P}\{S_{n-2} > (a-\gamma)y\}\right).$$

Therefore, the first term on the right-hand side of (4.2.19) is equivalent to

$$\text{(4.2.22)}\qquad \binom{n}{2}\cdot c_{\alpha_1}^2\cdot y^{-2\alpha_1}\cdot\mu(a,b)$$

as $n \to \infty$ with $y \geq (C n\cdot\log n)^{1/2}$.

Now, it is only left to show that the second sum on the right-hand side of (4.2.19) is negligible in comparison with expression (4.2.22). It is easily seen that this sum does not exceed

$$3\cdot\sum_{1\leq i<j\leq n}\mathbf{P}\{S_n - X_i - X_j > ay, X_i > \gamma y, X_j < -\gamma y\} + O\left(n^3\cdot y^{-3\alpha_1}\right)$$

$$= 3 \cdot \binom{n}{2} \cdot \mathbf{P}\{S_{n-2} > ay\} \cdot \mathbf{P}\{X_i > \gamma y\} \cdot \mathbf{P}\{X_j < -\gamma y\} + O\left(n^3 \cdot y^{-3\alpha_1}\right). \quad \square$$

Now, let us conclude the proof of Theorem 4.2.1 by the use of Lemmas 4.2.1 - 4.2.9. Note that all the following representations for $\{S_{n,y} \in G_{a,b}\}$ hold as $n \to \infty$, $y/\sqrt{n} \to \infty$.

a) We have that

$$\mathbf{P}\{S_{n,y} \in G_{a,b}\} = \Pi_1 + O(\Pi_2).$$

Applying Lemmas 4.2.1 - 4.2.3 we obtain the result that there exists a positive constant c such that the assertion of Theorem 4.2.1 holds in the range of deviations $y \le (c \cdot n \cdot \log n)^{1/2}$.

b) We have that

$$\mathbf{P}\{S_{n,y} \in G_{a,b}\} = \Pi_1 + \Pi_3 + O(\Pi_4) + \Pi_5 + O(\Pi_7).$$

Applying Lemmas 4.2.2 - 4.2.4, 4.2.6, 4.2.8 and 4.2.9 we obtain the result that for arbitrary $0 < c < C < \infty$ the assertion of Theorem 4.2.1 holds in the range of deviations $(c \cdot n \cdot \log n)^{1/2} \le y \le (C \cdot n \cdot \log n)^{1/2}$;

c) We have that

$$\mathbf{P}\{S_{n,y} \in G_{a,b}\} = \Pi_5 + O(\Pi_6) + O(\Pi_7) + O(\Pi_8).$$

Applying Lemmas 4.2.5 - 4.2.7 and 4.2.9 we obtain the result that there exists a positive constant C such that the assertion of Theorem 4.2.1 holds in the range of deviations $y \ge (C \cdot n \cdot \log n)^{1/2}$.

To complete the proof of Theorem 4.2.1 it only remains to combine points (a), (b) and (c). $\square$

Remark 4.2.2. It can be shown that in the case $b < a \cdot 2^{1/2}$, the middle term on the right-hand side of (4.2.3) drops being absorbed into the sum of the first and the third terms.

Remark 4.2.3. Let $\mathscr{E}_2$ be a subset of $\mathbf{D}[0,1]$ such that

$$\mathscr{E}_2 := \{f \in \mathbf{D}[0,1] : f(1) - \sup_{0<t\le 1} \left(f(t) - f(t-)\right) > a, f(1) > b,$$

$$\exists\ 0 < t_1 < t_2 \le 1 : f(t_i) - f(t_i -) > \gamma,\ i = 1,2\},$$

where γ is a fixed real belonging to the open interval $(0,(a/8) \wedge (r/40) \wedge (b-a)/2)$. Then it can be shown by the use of arguments similar to those used in the proof of Lemma 4.2.9 that

$$\mathbf{P}\{S_{n,y} \in \mathscr{E}_2\} \sim \binom{n}{2} \cdot c_{\alpha_1}^2 \cdot y^{-2\alpha_1} \cdot \mu(a,b)$$

as $n \to \infty$, $y/\sqrt{n} \to \infty$. Let us also note that the set $\mathscr{E}_2$ is neither w-continuous nor w-saturated, and r.s.f. $S_{n,y}$ hits the set $\mathscr{E}_2$ mainly due to the piecewise constant paths which perform two big jumps.

Now, let us demonstrate that for certain sets from $\mathbf{D}[0,1]$ the role of close-to-continuous paths may not be essential, whereas the role of piecewise constant paths performing two big jumps (by absolute value) and paths, which perform one big jump and are close to Wiener trajectories before and after the jump are essential. To this end, we introduce the following set:

$$B_{a,b} := \{f \in \mathbf{D}[0,1]: f(1) - \sup_{t \in [0,1]} \big(f(t) - f(t-)\big) < -a, f(1) > b\}, \tag{4.2.23}$$

where a and b are arbitrary positive reals (Note that the set $B_{a,b}$ is neither w-continuous nor w-saturated.) It is clear that

$$\mathbf{P}\{S_n - X_1(n) < -ay,\ S_n > by\} = \mathbf{P}\{S_{n,y} \in B_{a,b}\}. \tag{4.2.24}$$

Theorem 4.2.2. Let Conditions (0.1) and (0.1′) be fulfilled with $\alpha_1 > 2$, $\mathbf{E}X_i = 0$, and $\mathbf{V}X_i = 1$. Let a and b be arbitrary positive reals. Then

$$\begin{aligned}\mathbf{P}\{S_{n,y} \in B_{a,b}\} &= n \cdot c_{\alpha_1} \cdot (b+a)^{-\alpha_1} \cdot y^{-\alpha_1} \cdot \mathbf{P}\{\sqrt{n-1} \cdot w(1) < -ay\} \cdot (1 + o(1)) \\ &+ \binom{n}{2} \cdot c_{\alpha_1} \cdot d_{\alpha_1} \cdot y^{-2\alpha_1} \cdot \Big(\int_{-\infty}^{-a} (b-v)^{-\alpha_1} \cdot d|v|^{-\alpha_1} + o(1)\Big)\end{aligned} \tag{4.2.25}$$

as $n \to \infty$, $y/\sqrt{n} \to \infty$.

Remark 4.2.4. It can be shown that there exists a positive constant $C = C(a,b)$ such that for any real $\kappa > 0$, in the range of large deviations

$$y \le ((C-\kappa) n \log n)^{1/2}$$

the first term on the right-hand side of (4.2.25) makes the main contribution to the sum, and the event

$$\{S_{n,y} \in B_{a,b}\}$$

occurs mainly due to the paths presented in Fig. 4.

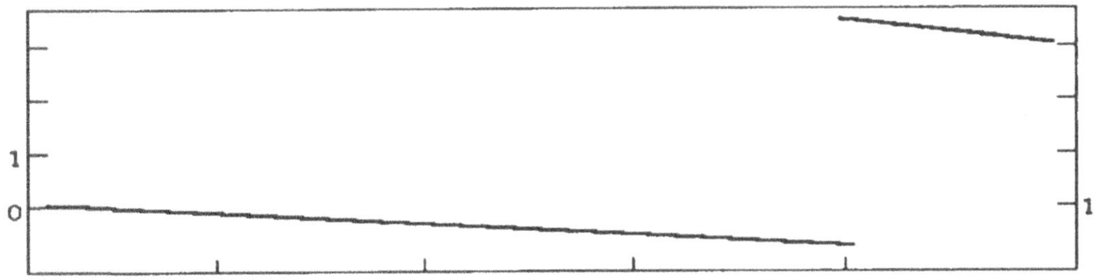

Figure 4

Namely, r.s.f. $S_{n,y}$ makes one big jump (at a random time instant uniformly distributed over the unit interval) with magnitude greater than $(b + a)$. In addition, if we glue the typical path of $S_{n,y}$ at the point of the maximal jump, then the modified trajectory of r.s.f. $S_{n,y}$ would resemble that of the Wiener process $\mathbf{w}(\cdot)$. Finally, if

$$y \geq ((C+\kappa)\cdot n\cdot \log n)^{1/2} ,$$

then the main contribution comes from the second term on the right-hand side of (4.2.25), and the most likely paths are piecewise constant and are presented in Fig. 5 and Fig. 6.

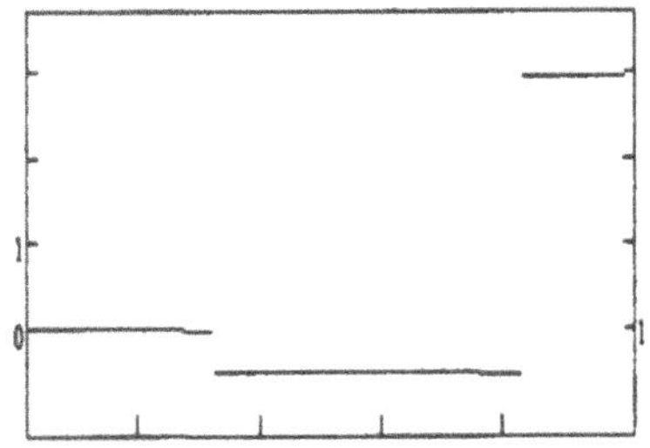

Figure 5

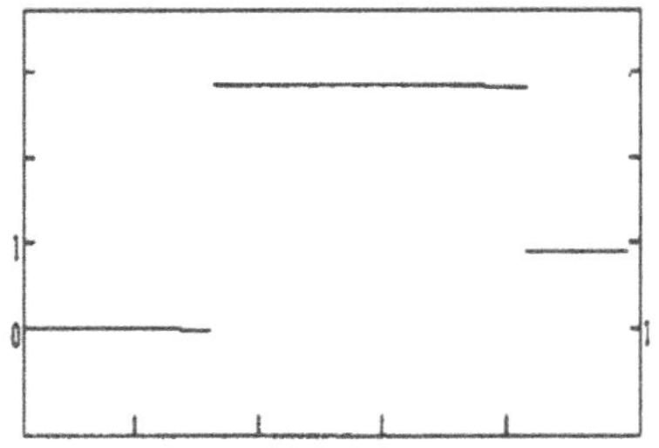

Figure 6

Proof of Theorem 4.2.2 basically follows along the same lines as that of Theorem 4.2.1 and therefore is omitted. Note, however, that the proof of Theorem 4.2.2 is much simpler than that of Theorem 4.2.1 because

$$\mathbf{P}\{n^{1/2}\cdot w(\cdot) \in y\cdot B_{a,b}\} = 0,$$

i.e., close-to-continuous realizations of r.s.f. $S_{n,y}(\cdot)$ do not make an essential contribution to the probability of the event $\{S_{n,y} \in B_{a,b}\}$. □

Now, in order to proceed with the next result, which describes the exact asymptotics of the probability of the following event:

(4.2.26) $\{S_n - X_1(n) - \dots - X_k(n) > y\}$,

we introduce some auxiliary notation (here $k \geq 1$ is any fixed integer, both n and y approach infinity). Let $\mathfrak{C} \subset (0,1]$ denote the set of all the time instants of jumps of the function f (which is either finite or countable set). If $\mathfrak{C}$ is countable then we introduce the quantities $\Delta_1 f, \dots, \Delta_k f$ as follows:

$$\Delta_1 f := \max_{s\in\mathfrak{C}} \big(f(s) - f(s-)\big);$$

$$\tau_1 := \min_{s\in\mathfrak{C}} \{s\colon f(s) - f(s-) = \Delta_1 f\};$$

$$\Delta_2 f := \max_{s\in\mathfrak{C}\setminus\tau_1} \big(f(s) - f(s-)\big);\ etc.$$

If $\mathfrak{C}$ is finite or empty (i.e. $Card\ \mathfrak{C} = N$) then $\Delta_1 f, \dots, \Delta_k f$ are defined as in the case of countable $\mathfrak{C}$, and $\Delta_{N+m} f := 0$ for any integer $m \geq 1$. Note that using the just introduced notation, we can rewrite the event (4.2.26) as $\{S_{n,y} \in \mathcal{F}_k\}$, where

(4.2.27) $\mathcal{F}_k := \{ f \in \mathbf{D}[0,1]\colon f(1) - \Delta_1 f - \dots - \Delta_k f > 1\}$.

Set

$$r := \rho(\mathcal{F}_k, \mathbf{B}^k).$$

Recall that ρ stands for the uniform metric on $\mathbf{D}[0,1]$. It can be shown that $r > 0$. In addition, note that the set $\mathcal{F}_k$ is w-continuous and w-saturated. The following result generalizes Theorem 2 of Vinogradov and Godovan'chuk (1989), where a much stronger Condition (0.1′) was assumed instead of Condition (0.10) that is posed below. In the next Theorem 4.2.3, we describe the exact asymptotics of the probability of r.s.f. $S_{n,y}$ to hit the set $\mathcal{F}_k$, in the full range of large deviations.

Theorem 4.2.3. Let Conditions (0.1) and (0.10) be fulfilled with $\alpha_1 > 2$, $\mathbf{E}X_i = 0$, and $\mathbf{V}X_i = 1$. Let $k \geq 1$ be any fixed integer. Then

$$\begin{aligned} \mathbf{P}\{S_{n,y} \in \mathcal{F}_k\} &= \mathbf{P}\{\sqrt{n}\cdot \mathbf{w}(1) > y\}\cdot\big(1 + o(1)\big) \\ &+ \binom{n}{k+1}\cdot c_{\alpha_1}^{k+1}\cdot y^{-(k+1)\alpha_1}\cdot\big(1 + o(1)\big) \end{aligned} \qquad (4.2.28)$$

as $n \to \infty$, $y/\sqrt{n} \to \infty$.

Remark 4.2.5. Relationship (4.2.28) naturally generalizes relationship (0.3) of the Introduction and refines Theorem 2.3 of Gafurov and Khamdamov (1987).

Proof of Theorem 4.2.3. Let $\beta = \beta(n)$ be the same as in the proof of Theorem 4.2.1. Fix an arbitrary $\gamma \in (0,1)$ and consider the following quantities which are similar to those introduced in the proof of Theorem 4.2.1:

$$\Pi_1 := \mathbf{P}\{S_{n,y} \in \mathscr{F}_k;\ |X_i| \le \beta n/y,\ i = 1, \dots , n\};$$

$$\Pi_2 = \mathbf{P}\Big(\bigcup_{i=1}^{n}\{|X_i| > \beta n/y\}\Big);$$

$$\Pi_3 = \sum_{t=1}^{n}\sum_{I_t}\mathbf{P}\{S_{n,y} \in \mathscr{F}_k;\ |X_i| > \beta n/y,\ i \in I_t,\ |X_j| \le \beta n/y,\ j \notin I_t\};$$

$$\Pi_4 := \sum_{I_{k+1}}\ \sum_{I_k \subset I_{k+1}} \mathbf{P}\{S_{n,y} \in \mathscr{F}_k;\ |X_i| > \beta n/y,\ i \in I_{k+1},\ |X_j| \le \gamma y,\ j \notin I_k\};$$

$$\Pi_5 := \sum_{I_{k+1}} \mathbf{P}\{S_{n,y} \in \mathscr{F}_k;\ |X_i| > \gamma y,\ i \in I_{k+1}\};$$

$$\Pi_6 := \sum_{t=1}^{k}\sum_{I_t} \mathbf{P}\{S_{n,y} \in \mathscr{F}_k;\ |X_i| > \gamma y,\ |X_j| \le \gamma y,\ j \notin I_t\};$$

$$\Pi_7 := \sum_{I_{k+2}} \mathbf{P}\{|X_i| > \gamma y,\ i \in I_{k+2}\};$$

$$\Pi_8 := \mathbf{P}\{S_{n,y} \in \mathscr{F}_k;\ |X_i| \le \gamma y,\ i = 1,\dots,n\}.$$

Here, the summation over I_t is understood as the summation over all arbitrary sets $\{i_1, \dots , i_t\}$ such that $1 \le i_1 < \dots < i_t \le n$.

By analogy with the proof of Theorem 4.2.1, we proceed with a series of lemmas which contain asymptotic representations for the quantities Π_i in various ranges of large deviations. Note that Lemmas 4.2.1 - 4.2.3 can be used without changes, whereas Lemmas 4.2.4 - 4.2.9 should be modified. We only formulate the analogs of these lemmas here (cf. Lemmas 4.2.4′ - 4.2.9′ below) skipping the proofs, since they are akin to the proofs of Lemmas 4.2.4 - 4.2.9.

Lemma 4.2.4′. Let $\beta(n)$ be a bounded positive sequence such that

$$\beta(n) := o(1/\log n)$$

as $n \to \infty$. Choose arbitrary $0 < c < C < \infty$. Then

$$\Pi_3 = o(\mathbf{P}\{n^{1/2} w(1) > y\})$$

as $n \to \infty$ with $(cn \cdot \log n)^{1/2} \le y \le (Cn \cdot \log n)^{1/2}$.

Lemma 4.2.5′. There exists a positive constant C such that

$$\Pi_8 = O\left(n^{k+2} \cdot y^{-(k+2)\alpha_1}\right)$$

as $n \to \infty$ with $y \ge (Cn \cdot \log n)^{1/2}$.

Lemma 4.2.6′. For any integer $n \ge 1$ and for any real $y > 0$,

$$\Pi_7 = O\left(n^{k+2} \cdot y^{-(k+2)\alpha_1}\right).$$

Lemma 4.2.7′. There exists a positive constant C such that

$$\Pi_6 = O\left(n^{k+2} \cdot y^{-(k+2)\alpha_1}\right)$$

as $n \to \infty$ with $y \ge (Cn \cdot \log n)^{1/2}$.

Lemma 4.2.8′. Choose arbitrary $0 < c < C < \infty$. Then

$$\Pi_4 = o\left(n^{k+1} \cdot y^{-(k+1)\alpha_1}\right)$$

as $n \to \infty$ with $(cn \cdot \log n)^{1/2} \le y \le (Cn \cdot \log n)^{1/2}$.

Lemma 4.2.9′. There exists a positive constant C such that

$$\Pi_5 \sim \binom{n}{k+1} \cdot c_{\alpha_1}^{k+1} \cdot y^{-(k+1)\cdot\alpha_1}$$

as $n \to \infty$ with $y \ge (Cn \cdot \log n)^{1/2}$.

To conclude the proof of Theorem 4.2.3, we apply Lemmas 4.2.1 - 4.2.3 and 4.2.4′ - 4.2.9′.

Note that all the following representations for $\mathbf{P}\{S_{n,y} \in \mathcal{F}_k\}$ hold as $n \to \infty$, $y/\sqrt{n} \to \infty$.

a) We have that

$$\mathbf{P}\{S_{n,y} \in \mathcal{F}_k\} = \Pi_1 + O(\Pi_2).$$

Applying Lemmas 4.2.1 - 4.2.3 we obtain the result that there exists a positive constant c such that the assertion of Theorem 4.2.3 holds in the range of deviations $y \le (cn \cdot \log n)^{1/2}$.

b) We have that

$$\mathbf{P}\{S_{n,y} \in \mathscr{F}_k\} = \Pi_1 + O(\Pi_3) + O(\Pi_4) + \Pi_5 + O(\Pi_7).$$

Applying Lemmas 4.2.2, 4.2.3, 4.2.4′, 4.2.6′, 4.2.8′ and 4.2.9′ we obtain the result that for arbitrary $0 < c < C < \infty$, the assertion of Theorem 4.2.3 holds in the range of deviations $(cn \cdot \log n)^{1/2} \le y \le (Cn \cdot \log n)^{1/2}$.

c) We have that

$$\mathbf{P}\{S_{n,y} \in \mathscr{F}_k\} = \Pi_5 + O(\Pi_6) + O(\Pi_7) + O(\Pi_8).$$

Applying Lemmas 4.2.5′ - 4.2.7′ and 4.2.9′ we obtain the result that there exists a positive constant C such that the assertion of Theorem 4.2.3 holds in the range of deviations $y \ge (Cn \cdot \log n)^{1/2}$.

To complete the proof of Theorem 4.2.3, it only remains to combine points (a), (b) and (c). □

Remark 4.2.6. It turns out that if Condition (0.1) is fulfilled with an arbitrary fixed positive index α_1 and $k \ge 1$ being an arbitrary fixed integer, then the following representation is valid (which can be viewed as a counterpart to Theorem 4.2.3 related to the case when the number n of summands in the sum S_n is **fixed**):

$$\mathbf{P}\{S_n - X_1(n) - \dots - X_k(n) > y\} \sim \binom{n}{k+1} \cdot y^{-(k+1)\alpha_1} \tag{4.2.29}$$

as $y \to \infty$.

Note that an analogous result for S_n with fixed n (which is due to Feller (1971) (Vol. II, Chapter 8, (8.14)) is quoted in the Introduction (see the result V therein).

Let us point out that one can also obtain the results analogous to Theorems 4.2.1 - 4.2.3 which describe the asymptotic behavior of the sum S_n without a fixed number of extremes determined by ordering the absolute values of the terms X_i. To be able to formulate the results of such kind, we introduce some auxiliary notation. Let

$$u_1 := \max_{1 \le i \le n} |X_i|, \qquad u_2 := \max_{1 \le i \le n,\, i \ne i_1} |X_i|,$$

$$i_1 := \min(i\colon |X_i| = u_1,\ 1 \le i \le n),\quad i_2 := \min(i\colon |X_i| = u_2,\ 1 \le i \le n, i \ne i_1)$$

$$\hat{X}_1(n) := X_{i_1}, \qquad\qquad \hat{X}_2(n) := X_{i_2}, \text{ etc.}$$

Set

$$\mathfrak{S}_{a,b} := \{f \in \mathbf{D}[0,1] : f(1) - \sup_{t\in[0,1]} |f(t) - f(t-)| > a, f(1) > b\}.$$

It is clear that

$$\mathbf{P}\{S_n - \hat{X}_1(n) > ay,\ S_n > by\} = \mathbf{P}\{S_{n,y} \in \mathfrak{S}_{a,b}\}.$$

It can be easily shown that $\rho(\mathfrak{S}_{a,b}, \mathbf{B}^1) > 0$. Let us also note that the sets $\mathfrak{S}_{a,b}$ are w-continuous and w-saturated. The next result was obtained in Godovan'chuk and Vinogradov (1990) Theorem 2′.

Theorem 4.2.4. Let Conditions (0.1) - (0.1′) be fulfilled with $\alpha_1 > 2$, $\mathbf{E}X_i = 0$, and $\mathbf{V}X_i = 1$. Let a and b be arbitrary fixed positive constants. Then

i) if $a > b$ then

$$\begin{aligned}\mathbf{P}\{S_{n,y} \in \mathfrak{S}_{a,b}\} &= \mathbf{P}\{\sqrt{n}\cdot w(1) > ay\}\cdot(1 + o(1)) \\ &+ \binom{n}{2}\cdot c_{\alpha_1}^2 \cdot a^{-2\alpha_1}\cdot y^{-2\alpha_1}\cdot(1 + o(1))\end{aligned} \tag{4.2.30}$$

as $n \to \infty$, $y/\sqrt{n} \to \infty$.

ii) If $a \le b$ then

$$\begin{aligned}\mathbf{P}\{S_{n,y} \in \mathfrak{S}_{a,b}\} &= \mathbf{P}\{\sqrt{n}\cdot w(1) > by\}\cdot(1 + o(1)) \\ &+ n\cdot c_{\alpha_1}\cdot(b-a)^{-\alpha_1}\cdot y^{-\alpha_1}\cdot \mathbf{P}\{\sqrt{n-1}\cdot w(1) > ay\}\cdot(1 + o(1)) \\ &+ \binom{n}{2}\cdot c_{\alpha_1}^2\cdot y^{-2\alpha_1}\cdot(\mu(a,b) + o(1))\end{aligned} \tag{4.2.31}$$

as $n \to \infty$, $y/\sqrt{n} \to \infty$, where $\mu(a,b)$ is given by (4.2.4).

Proof of Theorem 4.2.4. First, we represent the probability of r.s.f. $S_{n,y}$ to hit the set $\mathfrak{S}_{a,b}$ as

$$\begin{aligned}&\mathbf{P}\{S_n - X_1(n) > ay,\ S_n > by,\ |X_1(n)| \ge |X_n(n)|\} \\ &+ \mathbf{P}\{S_n - X_n(n) > ay,\ S_n > by,\ |X_1(n)| < |X_n(n)|\}.\end{aligned} \tag{4.2.32}$$

One can show that the second term of the sum (4.2.32) is negligible compared with the first one. The latter in turn is equivalent to

$$\mathbf{P}\{S_n - X_1(n) > ay, S_n > by\}$$

as $n \to \infty$, $y/\sqrt{n} \to \infty$. It is evident that if $a > b$ then the above probability is equal to $\mathbf{P}\{S_n - X_1(n) > ay\}$, and (4.2.30) of point (i) follows from Theorem 4.2.3 for the case $k = 1$. On the other hand, if $a \le b$ then (4.2.31) of point (ii) follows from Theorem 4.2.1. □

Now, we proceed with the conclusive result of this section (which is due to Godovan'chuk and Vinogradov (1990) Theorem 1.b) that describes the exact asymptotics of the probability of the following event:

(4.2.33) $\qquad \{S_n - \hat{X}_1(n) - \dots - \hat{X}_k(n) > y\}.$

Here, $k \ge 1$ is any fixed integer, and both n and y approach infinity.

Theorem 4.2.5. Let Conditions (0.1) and (0.1′) be fulfilled with $\alpha_1 > 2$, $\mathbf{E}X_i = 0$, and $\mathbf{V}X_i = 1$. Let $k \ge 1$ be any fixed integer. Then

(4.2.34)
$$\begin{aligned}\mathbf{P}\{S_n - \hat{X}_1(n) - \dots - \hat{X}_k(n) > y\} &= \mathbf{P}\{\sqrt{n}\cdot w(1) > y\}\cdot(1 + o(1)) \\ &+ \binom{n}{k+1}\cdot(c_{\alpha_1} + d_{\alpha_1})\cdot c_{\alpha_1}^k \cdot y^{-(k+1)\alpha_1}\cdot(1 + o(1))\end{aligned}$$

as $n \to \infty$, $y/\sqrt{n} \to \infty$.

Proof of Theorem 4.2.5 is analogous to that of Theorem 4.2.3. It involves an application of Lemmas 4.2.1 - 4.2.3 and modifications of Lemmas 4.2.4′ - 4.2.9′. □

Remark 4.2.7. The above Theorems 4.2.1 - 4.2.5 exhibit the variety of types of the asymptotics of the probabilities of large deviations of r.s.f. $S_{n,y}(\cdot)$. The problem of description of all the types of the asymptotics remains open.

Remark 4.2.8. Note that the study of the asymptotic behavior of trimmed sums in the case where the distributions of maxima belong to the domain of attraction of the Gumbel distribution is not of the same interest, since it is essentially similar to the study of the asymptotic behavior of S_n (cf., e.g., Lo (1989) for discussion).

4.3 Conditional Limit Theorems on Large Deviations for Trimmed Sums.

In this section, we derive a number of corollaries from the results of Section 4.2, which are related to the conditional asymptotics of the probabilities of large deviations of trimmed sums provided that the sum S_n also takes values in the range of large deviations. Let us point out that Corollaries 4.3.1 - 4.3.6 are comprised of two parts, which describe the asymptotics of conditional probabilities in the cases of power tails with index $\alpha_1 \in (0,1) \cup (1,2)$ and with index $\alpha_1 > 2$, respectively. Note that in the latter case, the results basically follow from the results of Section 4.2, whereas their counterparts related to the former case are derived from the results by Godovan'chuk (1978, 1981). In addition to this, we also employ the results I, II, and III from the Introduction (cf., e.g., formulas (0.2) and (0.3)). Points (i) and (ii) of Corollaries 4.3.1 - 4.3.6 are referred to the cases of powers tails with index $\alpha_1 \in (0,1) \cup (1,2)$, and with index $\alpha_1 > 2$, respectively. Note that preliminary versions of some results of this section were given in Vinogradov and Godovan'chuk (1989) and in Godovan'chuk and Vinogradov (1990).

Throughout this section, we assume the fulfilment of Condition (0.1) with index $\alpha_1 \in (0,1) \cup (1,2) \cup (2,\infty)$ In addition, in points (i) of Corollaries 4.3.1 - 4.3.6 (related to the case $\alpha_1 \in (0,1) \cup (1,2)$) we assume the fulfilment of Condition (0.1′). Alternatively, in point (ii) of Corollary 4.3.1 (related to the case $\alpha_1 > 2$) we assume the finiteness of a certain absolute moment of X_i's of order greater than two, whereas in Corollary 4.3.1′ we only assume the finiteness of the variance of X_i's. Consequently, in points (ii) of Corollaries 4.3.2 - 4.3.4 (also related to the case $\alpha_1 > 2$) we assume the fulfilment of Condition (0.10) with $\beta = \alpha_1$, whereas in points (ii) of Corollaries 4.3.5 - 4.3.6 we need to assume the fulfilment of Condition (0.1′). Finally, we assume without any loss of generality that $\mathbf{E}X_i = 0$ if $\alpha_1 > 1$ and $\mathbf{V}X_i = 1$ if $\alpha_1 > 2$.

Corollary 4.3.1 (compare to formula (0.15) which is due to Durrett (1980)).

i) Let $\alpha_1 < 2$. Then

(4.3.1) $$\mathbf{P}\{X_1(n) > y \mid S_n > y\} \to 1$$

as $n \to \infty$, $y/n^{1/\alpha_1} \to \infty$.

ii) Let $\alpha_1 > 2$. Then

$$(4.3.1') \qquad \mathbf{P}\{X_1(n) > y \mid S_n > y\} \sim \frac{n \cdot c_{\alpha_1} \cdot y^{-\alpha_1}}{n \cdot c_{\alpha_1} \cdot y^{-\alpha_1} + 1 - \Phi(y/\sqrt{n})}$$

as $n \to \infty$, $y/\sqrt{n} \to \infty$.

It should be mentioned that in order to derive (4.3.1′), we pose the assumption that $\mathbf{E}|X_i|^{2+\delta} < \infty$ for some $\delta > 0$, since this assumption is essential for fulfilment of the result III of the Introduction (see relationship (0.3) therein). However, this assumption could be weakened. Namely, if we replace this by the assumption that just the common variance $\mathbf{V}X_i$ of X_i's is finite then the following result is valid.

Corollary 4.3.1′. Let $\alpha_1 > 2$. Then

$$(4.3.1'') \qquad \mathbf{P}\{X_1(n) > y \mid S_n > y\} \to 1$$

as $n \to \infty$ with $y/\log y \geq \sqrt{n}$.

Corollary 4.3.2. Let $x \geq y$ and $k \geq 1$ be any fixed integer. Then

i) if $\alpha_1 < 2$ then

$$(4.3.2) \qquad \mathbf{P}\{S_n - X_1(n) - \ldots - X_k(n) > x \mid S_n > y\} \sim \frac{n^k \cdot c_{\alpha_1}^k}{(k+1)!} \cdot y^{\alpha_1} \cdot x^{-(k+1)\alpha_1}$$

as $n \to \infty$, $y/n^{1/\alpha_1} \to \infty$.

ii) Let $\alpha_1 > 2$. Then

$$\mathbf{P}\{S_n - X_1(n) - \ldots - X_k(n) > x \mid S_n > y\}$$

$$(4.3.2') \qquad \sim \frac{\binom{n}{k+1} \cdot c_{\alpha_1}^{k+1} \cdot y^{-(k+1)\alpha_1} + 1 - \Phi(x/\sqrt{n})}{n \cdot c_{\alpha_1} \cdot y^{-\alpha_1} + 1 - \Phi(y/\sqrt{n})}$$

as $n \to \infty$, $y/\sqrt{n} \to \infty$.

Now, let us consider the case when $x < y$. Then the following result is valid.

Corollary 4.3.3. Let $x = \theta \cdot y$ with constant θ belonging to the interval $[0,1]$. Then

i) if $\alpha_1 < 2$ then

$$(4.3.3) \qquad \mathbf{P}\{S_n - X_1(n) > x \mid S_n > y\} \sim 1/2 \cdot n \cdot c_{\alpha_1} \cdot y^{-\alpha_1} \cdot \mu(\theta,1)$$

as $n \to \infty$, $y/n^{1/\alpha_1} \to \infty$.

ii) if $\alpha_1 > 2$ then

$$\mathbf{P}\{S_n - X_1(n) > x \mid S_n > y\}$$

(4.3.3′)

$$\sim \left(\binom{n}{2} c_{\alpha_1}^2 y^{-2\alpha_1} \mu(\theta,1) + n c_{\alpha_1} (1-\theta)^{-\alpha_1} y^{-\alpha_1} \left(1 - \Phi(\theta y/\sqrt{n})\right) + 1 - \Phi(y/\sqrt{n})\right)$$
$$\cdot \left(n \cdot c_{\alpha_1} \cdot y^{-\alpha_1} + 1 - \Phi(y/\sqrt{n})\right)^{-1}$$

as $n \to \infty$, $y/\sqrt{n} \to \infty$. Here, the function $\mu(\cdot,\cdot)$ is defined by means of (4.2.4).

It is not difficult to get the following (slightly more general but less precise) result:

Corollary 4.3.4. Let $k \geq 2$ be any fixed integer, $x := \theta \cdot y$ with constant θ belonging to the interval $[0,1]$. Then

i) if $\alpha_1 < 2$ then

(4.3.4)

$$\mathbf{P}\{S_n - X_1(n) - \ldots - X_k(n) > x \mid S_n > y\} \sim \frac{n^k \cdot c_{\alpha_1}^k \cdot y^{-k\alpha_1}}{(k+1)!}$$
$$\cdot \int_{\substack{u_1 + \ldots + u_{k+1} > 1 \\ u_1 \wedge \ldots \wedge u_{k+1} > \theta}} \cdots \int d(-u_1^{-\alpha_1}) \cdot \ldots \cdot d(-u_{k+1}^{-\alpha_1})$$

as $n \to \infty$, $y/n^{1/\alpha_1} \to \infty$;

ii) if $\alpha_1 > 2$ then for any real $\kappa > 0$, representation (4.3.4) holds as $n \to \infty$ with $y \geq Const(\kappa)\, n^{1/2+\kappa}$.

The case of left-hand large deviations also deserves being studied. Here, we present only one result describing the conditional asymptotics of the probabilities of left-hand large deviations:

Corollary 4.3.5. Let $x = \theta \cdot y$, where θ is any fixed positive constant. Then

i) if $\alpha_1 < 2$ then

(4.3.5)
$$\mathbf{P}\{S_n - X_1(n) < -x \mid S_n > y\} \sim n \cdot d_{\alpha_1} \cdot y^{-\alpha_1} \cdot \int_{-\infty}^{-\theta} (1-u)^{-\alpha_1} \cdot d|u|^{-\alpha_1}$$

as $n \to \infty$, $y/n^{1/\alpha_1} \to \infty$.

ii) if $\alpha_1 > 2$ then

$$\begin{aligned}\mathbf{P}\{S_n - X_1(n) < -x \mid S_n > y\} \sim \Big(& nc_{\alpha_1}(1+\theta)^{-\alpha_1} y^{-\alpha_1} \Phi\Big(-\frac{\theta y}{\sqrt{n-1}}\Big) + 1 - \Phi\Big(\frac{y}{\sqrt{n}}\Big) \\ & + n^2 c_{\alpha_1} d_{\alpha_1} y^{-2\alpha_1} \cdot \int_{-\infty}^{-\theta} (1-u)^{-\alpha_1} \cdot d|u|^{-\alpha_1}\Big) \cdot \Big(nc_{\alpha_1} y^{-\alpha_1} + 1 - \Phi\Big(\frac{y}{\sqrt{n}}\Big)\Big)^{-1}\end{aligned} \tag{4.3.5'}$$

as $n \to \infty$, $y/\sqrt{n} \to \infty$.

Proofs of Corollaries 4.3.1 - 4.3.5 and 4.3.1′ are not at all difficult. All of these corollaries are easily derived from Theorems 4.2.1 - 4.2.3 of Section 4.2, the hereabove mentioned results by Godovan'chuk (1978, 1981), and the results I and II from the Introduction (see formulas (0.2) and (0.3) therein). We should also represent conditional probabilities as ratios of the corresponding probabilities of the intersection of pairs of events and probabilities of conditions. □

Note that one can also derive analogous results regarding the asymptotics of $\hat{X}_1(n)$, $S_n - \hat{X}_1(n)$, $S_n - \hat{X}_1(n) - \ldots - \hat{X}_k(n)$, etc. by applying Theorems 4.2.4 and 4.2.5 of the previous section. Here, we present only one result of such kind (that will be used in the proof of Theorems 4.4.1 and 4.4.2 of the next section).

Corollary 4.3.6.

$$\mathbf{P}\{|\hat{X}_1(n)| > y \mid |S_n| > y\} \to 1 \tag{4.3.6}$$

as $n \to \infty$, $y/n^{1/\alpha_1} \to \infty$ if $\alpha_1 < 2$ and $y/\log y \geq \sqrt{n}$ if $\alpha_1 > 2$.

Remark 4.3.1. It is clear that one can treat Corollary 4.3.6 as a two-sided counterpart of Corollary 4.3.1′.

4.4 Conditional Limit Theorems on Weak Convergence for Trimmed Sums.

In this section, we state and prove two conditioned limit theorems on weak convergence to stable laws for trimmed sums provided that the sum S_n takes values in the range of large deviations. It should be mentioned that in order to derive the results of this section, we employ Corollaries 4.3.1′ and 4.3.6 of the previous section. Let us also note that relationship (4.4.1) of Theorem 4.4.1 and relationship (4.4.11) of Theorem 4.4.2 were announced in Vinogradov and Godovan'chuk (1989) Theorems 5-6, but under stronger assumptions on the behavior of $F(\cdot)$.

Now, let us proceed with

Theorem 4.4.1. Let Conditions (0.1) be fulfilled with index $\alpha_1 \in (0,1) \cup (1,2) \cup (2,\infty)$. Let Condition (0.1′) be fulfilled if $\alpha_1 \in (0,1) \cup (1,2)$, or $\mathbf{V}X_i < \infty$ if $\alpha_1 > 2$. Let us also assume without loss of generality that $\mathbf{E}X_i = 0$ if $\alpha_1 > 1$. Let $\alpha := \alpha_1 \wedge 2$, and $B_n := n^{1/\alpha}$. Let $G_\alpha(\cdot)$ be the stable distribution function (limit of the random sequence S_n/B_n; G_2 is then the Laplace function Φ). Then

i)

(4.4.1) $\qquad \mathbf{P}\{(S_n - X_1(n))/B_{n-1} \le x \mid S_n > y\} \Rightarrow G_\alpha(x)$

as $n \to \infty$, $y/n^{1/\alpha} \to \infty$ if $\alpha_1 < 2$ and $y/\log y \ge \sqrt{n}$ if $\alpha_1 > 2$.

ii) If we assume the fulfilment of the additional Condition (0.1′) then

(4.4.1′) $\qquad \mathbf{P}\{(S_n - \hat{X}_1(n))/B_{n-1} \le x \mid S_n > y\} \Rightarrow G_\alpha(x)$

as $n \to \infty$, $y/n^{1/\alpha} \to \infty$ if $\alpha_1 < 2$ and $y/\log y \ge \sqrt{n}$ if $\alpha_1 > 2$.

Proof of Theorem 4.4.1.

i) In order to establish (4.4.1), we fix an arbitrary $x \in \mathbf{R}^1$. Obviously,

(4.4.2) $\qquad \mathbf{P}\{S_n - X_1(n) \le x \cdot B_{n-1} \mid S_n > y\}$

$$= 1 - \mathbf{P}\{S_n - X_1(n) > x \cdot B_{n-1}, S_n > y\} / \mathbf{P}\{S_n > y\}.$$

It is clear that the numerator of the latter fraction can be rewritten as

$$\mathbf{P}\{S_n > y \vee (x \cdot B_{n-1} + X_1(n))\} = \Pi_1 + \Pi_2,$$

where

$$\Pi_1 := \mathbf{P}\{S_n > y; X_1(n) \le y - x \cdot B_{n-1}\},$$

and

$$\Pi_2 := \mathbf{P}\{S_n - X_1(n) > x \cdot B_{n-1}, X_1(n) > y - x \cdot B_{n-1}\}.$$

It is easily seen that for $x \ge 0$,

$$\Pi_1 \le \mathbf{P}\{S_n > y, X_1(n) \le y\}.$$

A subsequent application of the result II from the Introduction and Corollary 4.3.1′ yields that

$$\mathbf{P}\{S_n > y, X_1(n) \le y\} = o(\mathbf{P}\{S_n > y\}).$$

Hence,

(4.4.3) $\Pi_1 = o(\mathbf{P}\{S_n > y\})$

for $x \geq 0$.

Now, let us show that formula (4.4.3) also remains valid for an arbitrary fixed $x < 0$. To this end, note that due to the already established fact that (4.4.3) is fulfilled for $x = 0$, we get that

$$\Pi_1 := \mathbf{P}\{S_n > y,\, X_1(n) \leq y - x\cdot B_{n-1}\}$$

(4.4.4) $$= \mathbf{P}\{S_n > y,\, y < X_1(n) \leq y - x\cdot B_{n-1}\} + o(\mathbf{P}\{S_n > y\}).$$

The first term on the right-hand side of (4.4.4) can be represented in the following form:

$$\mathbf{P}\{S_n > y,\, y < X_1(n) \leq y - x\cdot B_{n-1}\}$$

(4.4.5) $$= n\cdot\mathbf{P}\{S_n > y,\, y < X_n \leq y - x\cdot B_{n-1}\} + o(\mathbf{P}\{S_n > y\}).$$

Therefore, it only remains to estimate the fist term on the right-hand side of (4.4.5). It is clear that this term equals

$$n\cdot\int_{xB_{n-1}}^{\infty}\mathbf{P}\{y\vee(y-z) < X_n \leq y - x\cdot B_{n-1}\}\cdot dF_{S_{n-1}}(z)$$

(4.4.6) $$= n\cdot\int_{xB_{n-1}}^{0}\mathbf{P}\{y-z < X_n \leq y - x\cdot B_{n-1}\}\cdot dF_{S_{n-1}}(z)$$

$$+\, n\cdot\int_{0}^{\infty}\mathbf{P}\{y < X_n \leq y - x\cdot B_{n-1}\}\cdot dF_{S_{n-1}}(z).$$

Note that in view of (0.1), the first term on the right-hand side of (4.4.6) can be written out as

$$n\cdot\int_{xB_{n-1}}^{0}\left(c_{\alpha_1}\cdot(y-z)^{-\alpha_1} - c_{\alpha_1}\cdot(y-x\cdot B_{n-1})^{-\alpha_1} + o(y^{-\alpha_1})\right)\cdot dF_{S_{n-1}}(z),$$

where the remainder $o(y^{-\alpha_1})$ tends to 0 as $n \to \infty$ uniformly in the range of integration. Therefore, the above expression can in turn be rewritten as

$$n \cdot c_{\alpha_1} \cdot y^{-\alpha_1} \cdot \int_{xB_{n-1}}^{0} \left((1 - z/y)^{-\alpha_1} - (1 - x \cdot B_{n-1}/y)^{-\alpha_1}\right) \cdot dF_{S_{n-1}}(z) + o(n \cdot y^{-\alpha_1})$$

(4.4.7)

$$n \cdot c_{\alpha_1} \cdot y^{-\alpha_1} \cdot o(1) \cdot \mathbf{P}\{x \cdot B_{n-1} < S_{n-1} \le 0\} + o(n \cdot y^{-\alpha_1}) = o(n \cdot y^{-\alpha_1}).$$

Let us now estimate the second term on the right-hand side of (4.4.6). It is clear that this term is equal to

$$n \mathbf{P}\{y < X_n \le y - x B_{n-1}\} \cdot \mathbf{P}\{S_{n-1} \ge 0\}$$

(4.4.8) $$= n c_{\alpha} y^{-\alpha}(1 - (1 - x B_{n-1}/y)^{-\alpha} + o(1)) \cdot \mathbf{P}\{S_{n-1} \ge 0\} = o(y^{-\alpha}).$$

A subsequent combination of (4.4.4) - (4.4.8) implies that (4.4.3) is also valid for an arbitrary fixed $x < 0$.

Now, let us obtain the asymptotics of the main term Π_2. It is relatively easy to see that

$$\Pi_2 = n \cdot \mathbf{P}\{X_n > y - x \cdot B_{n-1}, S_{n-1} > x \cdot B_{n-1}, X_n > X_1(n-1)\} + O\left(n^2 \cdot y^{-2\alpha_1}\right)$$

$$= n \cdot \mathbf{P}\{X_n > y - x \cdot B_{n-1}, S_{n-1} > x \cdot B_{n-1}\} + O\left(n^2 \cdot y^{-2\alpha_1}\right)$$

(4.4.9)

$$- n \cdot \mathbf{P}\{X_n > y - x \cdot B_{n-1}, S_{n-1} > x \cdot B_{n-1}, X_n \le X_1(n-1)\}$$

$$= n \cdot c_{\alpha_1} \cdot (y - x \cdot B_{n-1})^{-\alpha_1} \cdot \mathbf{P}\{S_{n-1} > x \cdot B_{n-1}\} \cdot \left(1 + o(1)\right) + O\left(n^2 \cdot y^{-2\alpha_1}\right).$$

Combining (4.4.2), (4.4.3) and (4.4.9) with the result II of the Introduction (see (0.2)) we obtain that

$$\mathbf{P}\{(S_n - X_1(n))/B_{n-1} \le x \mid S_n > y\} = 1 - n c_{\alpha} (y - x B_{n-1})^{-\alpha}$$

(4.4.10) $$\cdot \; \mathbf{P}\{S_{n-1}/B_{n-1} > x\} \,/\, \mathbf{P}\{S_n > y\} + o(1) \sim \mathbf{P}\{S_{n-1}/B_{n-1} \le x\}$$

as $n \to \infty$ for any fixed real x.

It only remains to apply the well-known theorems on weak convergence of the distribution function of r.v. S_{n-1}/B_{n-1} to $G_\alpha(\cdot)$ as $n \to \infty$. This completes the proof of (4.4.1). □

The proof of (4.4.1′) is analogous to that of (4.4.1). The difference is that one should apply Corollary 4.3.6 (instead of Corollary 4.3.1′) in order to establish that

$$\mathbf{P}\{\,|S_n| > y,\ |\hat{X}_1(n)| \le y - x B_{n-1}\} = o(\mathbf{P}\{\,|S_n| > y\})$$

for any fixed real x, and as $n \to \infty$, $y/n^{1/\alpha} \to \infty$ if $\alpha_1 < 2$, and with $y/\log y \ge \sqrt{n}$ if $\alpha_1 > 2$. □

The following result reveals 'conditional' weak limits of trimmed sums in the full range of large deviations of maxima.

Theorem 4.4.2. Assume that all the conditions of Theorem 4.4.1 are fulfilled. Then

i)

(4.4.11) $$\mathbf{P}\{(S_n - X_1(n))/B_{n-1} \le x \mid X_1(n) > y\} \Rightarrow G_\alpha(x)$$

as $n \to \infty$, $y/n^{1/\alpha} \to \infty$.

ii)

(4.4.11′) $$\mathbf{P}\{(S_n - \hat{X}_1(n))/B_{n-1} \le x \mid \hat{X}_1(n) > y\} \Rightarrow G_\alpha(x)$$

as $n \to \infty$, $y/n^{1/\alpha} \to \infty$.

Proof of Theorem 4.4.2 is analogous to that of Theorem 4.4.1 and therefore is omitted. □

Remark 4.4.1. Note that formulas (4.4.1) - (4.4.1′) and (4.4.11) - (4.4.11′) demonstrate that the classical stable distributions emerge as weak limits in *conditioned limit theorems on weak convergence* for trimmed sums, whereas *unconditioned* weak limits of trimmed sums are different (cf., e.g., the works by Arov and Bobrov (1960), Hall (1978), Teugels (1981), Csörgő, Csörgő, Horváth and Mason (1986) quoted in the Introduction).

Chapter 5

Large Deviations for I.I.D. Random Sums when Cramér's Condition is Fulfilled Only on a Finite Interval.

5.1 Exact Asymptotics of the Probabilities of Large Deviations and of the Expectations of Smooth Functions of I.I.D. Random Sums in the Case of Exponential-Power Tails.

First of all, let us present the following lemma, which reformulates Condition (0.16) on the asymptotic behavior of the right-hand tail of the given distribution $F(\cdot)$ in terms of the asymptotic behavior of the right-hand tail of the transformed distribution $F^{z_0}(\cdot)$.

Lemma 5.1.1. Let Condition (0.16) be fulfilled with $a = a_1$. Then

i) The distribution function $F(\cdot)$ is not lattice.

ii) Let $F^{z_0}(\cdot)$ denote the common d.f. of i.i.d.r.v.'s $\{X_n,\ n \geq 1\}$ with respect to μ^{z_0}. Then

(ii1) $$1 - F^{z_0}(y) \sim \frac{z_0 \cdot C_{a_1}}{\psi(z_0) \cdot (a_1 - 1)} \cdot y^{-(a_1-1)}$$

as $y \to \infty$.

(ii2) $$|F^{z_0}(y+x) - F^{z_0}(y)| \sim \frac{z_0 \cdot C_{a_1}}{\psi(z_0)} \cdot |x| \cdot y^{-a_1}$$

as $y \to \infty$, where $0 < Const \leq |x| = o(y)$.

Proof of Lemma 5.1.1. Point i) is carried out by contradiction. To this end, assume that the distribution $F(\cdot)$ is spanned by the lattice $\{b + d \cdot h\}$ with the grid $h > 0$, $d \in \mathbf{Z}$ and $b \in \mathbf{R}^1$. Then (0.16) implies that

$$F(b+(d+1)\cdot h) - F((b+(d+1)\cdot h)-) = o(\exp\{-z_0 \cdot (b+(d+1)\cdot h)\} \cdot (b+(d+1)\cdot h)^{-a_1}).$$

On the other hand, the fact that the distribution F is not lattice implies that

$$F(b+(d+1)\cdot h) - F((b+(d+1)\cdot h)-) = F(b+(d+1)\cdot h) - F(b+d\cdot h)$$
$$= O(\exp\{-z_0 \cdot (b+(d+1)\cdot h)\} \cdot (b+(d+1)\cdot h)^{-a_1} \cdot (\exp\{z_0 \cdot h\} - 1)),$$

a contradiction. □

The **proof of ii1)** is simple and therefore is omitted. It essentially relies on the integration by parts. □

The **proof of ii2)** is split into two parts, which correspond to the sign of the parameter x. Let us assume for simplicity that $x > 0$. Then we should show that

$$F^{z_0}(y + x) - F^{z_0}(y) = \mathbf{P}^{z_0}\{y < X_i \leq y + x\}$$

$$\sim \psi(z_0)^{-1}\cdot(a_1 - 1)^{-1}\cdot z_0\cdot C_{a_1}\cdot x\cdot y^{-a_1}. \tag{5.1.1}$$

By definition,

$$\mathbf{P}^{z_0}\{y < X_i \leq y + x\} = \psi(z_0)^{-1}\cdot\int_y^{y+x}\exp\{z_0 v\}\cdot d(F(v) - 1)$$

$$= \psi(z_0)^{-1}\cdot\Big\{-\exp\{z_0\cdot y\}\cdot\big(F(y) - 1\big) + \exp\{z_0\cdot(y+x)\}\cdot\big(F(y+x) - 1\big)$$

$$+ z_0\cdot\int_y^{y+x}\big(1 - F(v)\big)\cdot\exp\{z_0\cdot v\}\cdot dv\Big\}.$$

Note that the difference of the values of the term outside the integral presented within the braces is equal to

$$C_{a_1}\cdot\big(y^{-a_1} - (y+x)^{-a_1}\big) + o(y^{-a_1}).$$

The latter expression can be rewritten as

$$O(x\cdot y^{-a_1-1}) + o(y^{-a_1}) = o(x\cdot y^{-a_1}). \tag{5.1.2}$$

Now, let us evaluate the integral presented within the braces. In view of (0.16), this integral is equal to

$$\int_y^{y+x}\big(C_{a_1}\cdot v^{-a_1} + o(v^{-a_1})\big)\cdot dv = -\frac{C_{a_1}}{a_1 - 1}\cdot v^{1-a_1}\Big|_y^{y+x} + o(x\cdot y^{-a_1})$$

$$= \frac{C_{a_1}}{a_1 - 1}\cdot\big(y^{-(a_1-1)} - (y+x)^{-(a_1-1)}\big) + o(x\cdot y^{-a_1}) \tag{5.1.3}$$

$$= C_{a_1}\cdot x\cdot y^{-a_1} + o(x\cdot y^{-a_1}).$$

A subsequent combination of (5.1.2) and (5.1.3) implies (5.1.1), and the assertion (ii2) is proved for the case of $x > 0$. The consideration of the case of $x < 0$ follows along the same lines as that for the case of x > 0 and therefore is omitted. □

Remark 5.1.1. Let us point out that ii1) implies that under fulfilment of (0.16), the asymptotics of the right-hand tail of the transformed distribution $F^{z_0}(\cdot)$ is of power type with index $\alpha_1 = a_1 - 1 > 1$. In addition, it is clear that the left-hand tail of the transformed distribution $F^{z_0}(\cdot)$ decays exponentially as $x \to -\infty$. Therefore, $\mathbf{E}_{z_0}X_1$ ($= u_0$) $< \infty$.

Now, let us introduce a stronger condition than Condition (0.16):

$$(5.1.4) \qquad 1 - F(x) = \sum_{i=1}^{t} C_{a_i}\cdot \exp\{-z_0 x\} x^{-a_i} + o\big(\exp\{-z_0 x\}\cdot x^{-s}\big)$$

as $x \to \infty$, where $2 < a_1 < a_2 < \ldots < a_t \le s$, and $s \ge a_1 + 1$.

The next lemma refines the assertion of Lemma 5.1.1ii1 for a special case of fulfilment of Condition (5.1.4):

Lemma 5.1.2. Let Condition (5.1.4) be fulfilled. Then

$$(5.1.5) \qquad 1 - F^{z_0}(y) = \psi(z_0)^{-1}\cdot\left(\sum_{i=1}^{t}\left(\frac{z_0\cdot C_{a_i}}{a_i - 1}\cdot y^{-(a_i-1)} + C_{a_i}\cdot y^{-a_i}\right)\right) + o(y^{-(s-1)})$$

as $x \to \infty$.

Proof of Lemma 5.1.2 is simple and is based on the integration by parts. □

Remark 5.1.2. It is clear that (5.1.5) implies that under fulfilment of (5.1.4), the right-hand tail of the transformed distribution $F^{z_0}(\cdot)$ admits an asymptotic expansion over negative powers of x as $x \to \infty$. This fact will be used in the proof of Theorem 5.1.1.

Now, let us proceed with the main result of this section, that generalizes the theorem of Vinogradov (1983).

Theorem 5.1.1 (compare to Vinogradov (1983, 1993), A.V. Nagaev (1985, 1989)). Let Condition (5.1.4) be fulfilled with $a = a_1 \in (2,3) \cup (3,\infty)$. Let $\kappa > 0$ be any fixed real. Then

(5.1.6) $$\mathbf{P}\{S_n > y\} \sim n\cdot\mathbf{P}\{X_1 > y - (n-1)\cdot u_0\}\cdot\exp\{-(n-1)\cdot\mathbf{H}(u_0)\}$$

as $n \to \infty$, $(y-nu_0)/n^{1/(a_1-2)} \to \infty$ if $a_1 \in (2,3)$ and $y - nu_0 \geq Const(\kappa)\cdot n^{1/2+\kappa}$ if $a_1 \in (3,\infty)$. Here, $\mathbf{H}(\cdot)$ denotes the rate function defined on page 21 of the Intoroduction.

Proof of Theorem 5.1.1 involves the Cramér's transformation with the subsequent application of Lemmas 5.1.1i and 5.1.2, Remark 5.1.1, Theorem 1.3.1 with $\alpha_1 = a_1 - 1$, $\beta = \alpha_1$, $r = a_1$ (in the case $2 < a_1 < 3$), and Theorem 1.3.2 with $\alpha_1 = a_1 - 1$, $\beta = 3$, $r = a_1$ (in the case $a_1 > 3$). Indeed, by (0.19′),

(5.1.7) $$\mathbf{P}\{S_n > y\} = \psi(z_0)^n \cdot \int_y^\infty \exp\{-z_0\cdot v\}\cdot d\left(F_{S_n}^{z_0}(v) - 1\right).$$

We approximate the expression under the sign of differential in (5.1.7) by the use of Theorem 1.3.1 if $2 < a_1 < 3$ and Theorem 1.3.2 if $a_1 > 3$. Clearly, the condition that the distribution $F^{z_0}(\cdot)$ is not lattice (requested in Theorem 1.3.2) easily follows from Lemma 5.1.1i. For simplicity, let us carry out the proof for the case of $2 < a_1 < 3$ only. Note that the proof for the case of $a_1 > 3$ is similar to the proof related to the case of $2 < a_1 < 3$ and can be found in Vinogradov (1983).

Now, note that $\mathbf{E}_n^{z_0}S_n = nu_0$. Therefore, an application of Theorem 1.3.1 implies that for $v \geq y$,

(5.1.8) $$\begin{aligned} F_{S_n}^{z_0}(v) - 1 &= -P_n^{z_0}\{S_n > v\} = -P_n^{z_0}\{S_n - nu_0 > v - nu_0\} \\ &= -n\cdot\psi(z_0)^{-1}\Bigg(\sum_{i=1}^{l}\left(\frac{z_0\cdot C_{a_i}}{a_i - 1}\cdot\left(v-(n-1)\cdot u_0\right)^{-(a_i-1)} + C_{a_i}\cdot\left(v-(n-1)\cdot u_0\right)^{-a_i}\right) \\ &\quad + O\left(n\cdot\left(v-(n-1)\cdot u_0\right)^{-2\cdot(a_1-1)}\right) + o\left(\left(v-(n-1)\cdot u_0\right)^{-(a-1)}\right)\Bigg) \end{aligned}$$

as $n \to \infty$, $(y-nu_0)/n^{1/(a_1-1)} \to \infty$. In addition, it is easily seen that if $y/n^{1/(a_1-2)} \to \infty$ as $n \to \infty$ then both remainders on the right-hand side of (5.1.8) can be replaced by

(5.1.9) $$o((v-(n-1)\cdot u_0)^{-a_1}).$$

Now, let us make some transformations with the integral on the right-hand side of (5.1.7).

Replace the expression under the sign of differential by the expression on the right-hand side of (5.1.8). Integrating by parts, we ascertain that

$$
\begin{aligned}
&\psi(z_0)^n \cdot \int_y^\infty \exp\{-z_0 \cdot v\} \cdot d\Big(o\big(n \cdot (v-(n-1)\cdot u_0)^{-a_1}\big)\Big) \\
(5.1.10)\qquad &= o\big(n \cdot \psi(z_0)^n \cdot (y-(n-1)\cdot u_0)^{-a_1}\big).
\end{aligned}
$$

Let us now evaluate the contribution to the probability $\mathbf{P}\{S_n > y\}$ that comes from the main, first term of the expansion on the right-hand side of (5.1.8). Namely, making the change of variables $w = v - (n-1)\cdot u_0$ we get that the contribution of this term is as follows:

$$
\begin{aligned}
&\psi(z_0)^n \cdot \int_y^\infty \exp\{-z_0 \cdot v\} \cdot d\left(-\frac{n}{\psi(z_0)} \cdot \frac{z_0 \cdot C_{a_1}}{a_1 - 1} \cdot (v-(n-1)\cdot u_0)^{-(a_1-1)}\right) \\
(5.1.11)\qquad &= n \cdot \psi(z_0)^{n-1} \cdot \exp\{-(n-1)\cdot z_0 \cdot u_0\} \cdot \frac{z_0 \cdot C_{a_1}}{a_1 - 1} \cdot \int_{y-(n-1)\cdot u_0}^\infty \exp\{-z_0 \cdot w\} \cdot d(-w^{-(a_1-1)})
\end{aligned}
$$

(by the Laplace method)

$$
\sim n \cdot \exp\{-(n-1)\cdot \mathbf{H}(u_0)\} \cdot \exp\{-z_0 \cdot (y-(n-1)\cdot u_0)\} \cdot C_{a_1} \cdot (y-(n-1)\cdot u_0)^{-a_1}
$$

(by Condition (5.1.4))

$$
\sim n \cdot \mathbf{P}\{X_1 > y - (n-1)\cdot u_0) \cdot \exp\{-(n-1)\cdot \mathbf{H}(u_0)\}.
$$

Moreover, it can be shown that the contribution of all the refining terms presented on the right-hand side of (5.1.8) is included to the remainder (5.1.9), multiplied by n. A combination of (5.1.7) - (5.1.11) implies (5.1.6). □

Remark 5.1.3. It should be pointed out that the assumptions of Theorem 5.1.1 can be weakened. Here, we decided to pose some extra conditions on the reason that the role of Theorem 5.1.1 in the present monograph is of an illustrative character. In particular, Theorem 5.1.1 demonstrates the range of applications of the results of Chapter 1. In addition, the introduction of some extra assumptions made it possible to simplify the proof in an essential way.

Remark 5.1.4. Let us underscore that under fulfilment of (5.1.4) and in the range of deviations $y \le n \cdot u_0$, the exact asymptotics of the probabilities of large deviations of S_n is given in terms of Cramér's formula. In particular, if $y = n \cdot u$ with $\mathbf{E}X_i + \varepsilon \le u \le u_0$, where $\varepsilon > 0$ is an arbitrary fixed real, then

$$(5.1.12) \qquad \mathbf{P}\{S_n > n \cdot u\} \sim \frac{\exp\{-n \cdot \mathbf{H}(u)\}}{z(u) \cdot \sqrt{2\pi n \cdot V''(z(u))}}$$

as $n \to \infty$ (cf., e.g., Petrov (1965) and A.V. Nagaev (1989)).

Remark 5.1.5. It is interesting to note that in the case of *fixed n*, Chover, Ney, and Wainger (1973) (see formula (2.9) therein) obtained a representation similar to our representation (5.1.6), which was in fact the key tool in deriving their conditioned limit theorems for subcritical branching processes. The relationship between the above formula (5.1.6) and formula (2.9) of Chover, Ney, and Wainger (1973) is similar to that between the results I and V of the Introduction.

The local analog of (5.1.6) is also true.

Theorem 5.1.2. Let i.i.d.r.v.'s $\{X_n, n \ge 1\}$ have the common bounded density $\mathbf{p}(\cdot)$ such that

$$(5.1.13) \qquad \mathbf{p}(x) \sim c_a \exp\{-z_0 x\} x^{-a}$$

as $x \to \infty$, where $z_0 > 0$, $c_a > 0$, and $a > 2$. Let $\kappa > 0$ be any fixed real. Then the density $\mathbf{p}_{S_n}(\cdot)$ of S_n has the following asymptotics on the right-hand tail:

$$\mathbf{p}_{S_n}(y) \sim n \cdot \mathbf{p}(y - (n-1) \cdot u_0) \cdot \exp\{-(n-1) \cdot \mathbf{H}(u_0)\}$$

as $n \to \infty$ with $y - n \cdot u_0 \ge Const(\kappa) n^{1/2+\kappa}$.

Proof of Theorem 5.1.2 is simple. It is based on an application of the Cramér's transformation and the limit theorems on large deviations for densities, in the case of power tails (obtained in A.V. Nagaev (1970)). □

Remark 5.1.6. Note that (5.1.13) implies (0.16) with the same a and $C_a = c_a/z_0$. In addition, applying arguments similar to those used in the proof of Theorem 5.1.1, one can show that relationship (5.1.6) also remains valid under fulfilment of the conditions of Theorem 5.1.2.

The asymptotics of expectations of functionals of S_n/n also deserves being studied. The next result was formulated in Vinogradov (1993) Theorem 2.4.

Theorem 5.1.3. Let Condition (0.16) be fulfilled with $a \in (2,3) \cup (3,\infty)$. Let $f \in \mathbf{C}^2(\mathbf{R}_+^1)$ be such that the function $z_0 \cdot \mathrm{v} - f(\mathrm{v})$ attains minimum at the point $\mathrm{v} = u_1 > u_0$ such that $f''(u_1) < 0$. Then

$$
\begin{aligned}
(5.1.14) \qquad & \mathbf{E}\exp\{n \cdot f(S_n/n)\} \sim \exp\{n \cdot (f(u_1) - \mathbf{H}(u_1))\} \\
& \cdot \sqrt{2\pi / |f''(u_1)|} \cdot \psi(z_0)^{-1} \cdot \frac{z_0 \cdot C_a}{a-1} \cdot u_0^{-a} \cdot n^{3/2-a}
\end{aligned}
$$

as $n \to \infty$.

Proof of Theorem 5.1.3 is carried out by a combination of the Laplace method, some arguments used in the proof of Theorem 5.1.1, Lemma 5.1.1, and the interval limit theorems on large deviations in the case of power tails derived in Tkachuk (1973) for $\alpha = a - 1 \in (1,2)$ and in A.V. Nagaev (1969) for $\alpha = a - 1 \in (2,\infty)$.

Recall that $u_0 < \infty$, then choose u_2 and u_3 such that $u_0 < u_2 < u_1 < u_3$. Then an application of (0.19′) implies that

$$
\begin{aligned}
(5.1.15) \qquad & \mathbf{E}\exp\{n \cdot f(S_n/n)\} = \exp\{n \cdot V(z_0)\} \cdot \mathbf{E}_n^{z_0}(\exp\{-z_0 \cdot S_n + n \cdot f(S_n/n)\}) \\
& = \exp\{n \cdot V(z_0)\} \cdot \int_{-\infty}^{\infty} \exp\{-n \cdot (z_0 y/n - f(y/n))\} \cdot dF_{S_n}^{z_0}(y).
\end{aligned}
$$

Now, let us split the integral on the right-hand side of (5.1.15) into three parts: from $-\infty$ to nu_2, from nu_2 to nu_3, and from nu_3 to ∞. Let us keep in mind our choice of u_2 and u_3 such that $u_0 < u_2 < u_1 < u_3$, along with the facts that the function $z_0 \cdot \mathrm{v} - f(\mathrm{v})$ attains minimum at the point $\mathrm{v} = u_1 > u_0$, and that for $u \geq u_0$, the rate function grows linearly: $\mathbf{H}(u) = \mathbf{H}(u_0) + z_0 \cdot (u - u_0)$. Then we get that both the first integral (from $-\infty$ to nu_2) and the third integral (from nu_3 to ∞), multiplied by $\exp\{n \cdot V(z_0)\}$, admit the following estimate: for any $\varepsilon > 0$, there exists an integer $N = N(\varepsilon)$ such that for each $n \geq N(\varepsilon)$, each of these expressions does not exceed

$$
(5.1.16) \qquad \exp\{n \cdot (f(u_1) - \mathbf{H}(u_1) - \varepsilon)\}.
$$

Let us evaluate the following expression which determines the asymptotics of the expectation on the left-hand side of (5.1.14):

$$(5.1.17) \qquad \exp\{n\cdot V(z_0)\}\cdot\int_{n\cdot u_2}^{n\cdot u_3}\exp\{-n\cdot(z_0 y/n - f(y/n))\}\cdot dF_{S_n}^{z_0}(y).$$

Making the change of variables $w = (y - nu_1)/\sqrt{n}$ we represent expression (5.1.17) in the following form:

$$\exp\{n\cdot V(z_0)\}\cdot\int_{-\sqrt{n}(u_1-u_2)}^{\sqrt{n}(u_3-u_1)}\exp\{-z_0 nu_1 - z_0\sqrt{n}w + n\cdot f(u_1 + w/\sqrt{n})\}$$

$$\cdot\, d(F_{S_n}^{z_0}(nu_1+\sqrt{n}\cdot w) - F_{S_n}^{z_0}(nu_1)).$$

Keeping in mind that for $u \geq u_0$, the rate function

$$\mathbf{H}(u) = \mathbf{H}(u) + z_0(u - u_0) = z_0 u - V(z_0)$$

and integrating the above integral by parts, we ascertain that the values taken by the term outside the integral at the lower and the upper limits are as follows:

$$(5.1.18) \qquad \begin{aligned}&\exp\{-n\cdot\mathbf{H}(u_1)\}\cdot\left(F_{S_n}^{z_0}(nu_2) - F_{S_n}^{z_0}(nu_1)\right)\cdot\exp\{-z_0 n(u_1-u_2) + n\cdot f(u_2)\}\\ &= O\left(\exp\left\{n\cdot\left(f(u_2) - \mathbf{H}(u_2)\right)\right\}\right),\end{aligned}$$

and

$$(5.1.18') \qquad \begin{aligned}&\exp\{-n\cdot\mathbf{H}(u_1)\}\cdot\left(F_{S_n}^{z_0}(nu_3) - F_{S_n}^{z_0}(nu_1)\right)\cdot\exp\{-z_0 n(u_3-u_1) + n\cdot f(u_3)\}\\ &= O\left(\exp\left\{n\cdot\left(f(u_3) - \mathbf{H}(u_3)\right)\right\}\right).\end{aligned}$$

It is clear that there exist a real $\varepsilon > 0$ and an integer $N = N(\varepsilon)$ such that for each $n \geq N$, both expressions (5.1.18) and (5.1.18′) are of the following order:

$$(5.1.19) \qquad O(\exp\{n\cdot(f(u_1) - \mathbf{H}(u_1) - \varepsilon)\}).$$

Therefore, it only remains to evaluate

$$
\begin{aligned}
&-\exp\{-n\cdot\mathbf{H}(u_1)\}\cdot\Bigg\{\int_0^{\sqrt{n}\cdot(u_3-u_1)} \mathbf{P}_n^{z_0}\{nu_1 < S_n \le nu_1 + \sqrt{n}\cdot w\} \\
&\quad\cdot d\Big(\exp\{-z_0\sqrt{n}w + n\cdot f(u_1 + w/\sqrt{n})\}\Big) \\
(5.1.20)\qquad &\quad - \int_{-\sqrt{n}\cdot(u_1-u_2)}^{0} \mathbf{P}_n^{z_0}\{nu_1 + \sqrt{n}\cdot w < S_n \le nu_1\} \\
&\quad\cdot d\Big(\exp\{-z_0\sqrt{n}w + n\cdot f(u_1 + w/\sqrt{n})\}\Big)\Bigg\}.
\end{aligned}
$$

Note that in order to evaluate both integrals within the brackets, we can use the assertion ii2) of Lemma 5.1.1, but only for such values of w that

$$0 < Const/\sqrt{n} \le |w| \le \sqrt{n}/\delta(n),$$

where $\delta(n)$ is a certain sequence of positive numbers that converges to zero as $n \to \infty$ with an arbitrary speed. In particular, one can choose this speed to be as slow as necessary. Therefore, we need to estimate the integrals from $-C_1/\sqrt{n}$ to zero and from zero to $C_2/\sqrt{n}$ by a different method. Note that both integrals presented in (5.1.20) are evaluated following along the same lines. Therefore, it suffices to estimate only the integral over the range $w \in [0, C_2/\sqrt{n}]$. It is easily seen that for $|w| \le Const/\sqrt{n}$,

$$
\begin{aligned}
&- z_0\sqrt{n}w + nf(u_1 + w/\sqrt{n}) = -z_0\sqrt{n}w + nf(u_1) \\
(5.1.21)\qquad &+ \sqrt{n}f'(u_1)\cdot w - \theta\cdot|f''(u_1)|\cdot w^2,
\end{aligned}
$$

where θ is a certain positive constant. Now, recall that the function f belongs to $\mathbf{C}^2(\mathbb{R}_+^1)$, and that the function $z_0 v - f(v)$ attains minimum at the point $v = u_1 > u_0$, such that $f''(u_1) < 0$. Hence, $f'(u_1) = z_0$, and the expression on the right-hand side of (5.1.21) can be rewritten as

$$(5.1.21') \qquad nf(u_1) - \theta\cdot|f''(u_1)|\cdot w^2.$$

In addition, note that it follows from the *interval limit theorems on large deviations* in the case of *power tails* derived in Tkachuk (1973) and in A.V. Nagaev (1969) that

$$
\text{(5.1.22)} \quad \begin{aligned} &\mathbf{P}_n^{z_0}\{nu_1 < S_n \le nu_1 + \sqrt{n}\cdot w\} \\ &\sim n\cdot\psi(z_0)^{-1}\cdot z_0\cdot C_a\cdot\sqrt{n}\cdot w\cdot(nu_1)^{-a} \end{aligned}
$$

as $n \to \infty$ with $0 < Const/\sqrt{n} \le |w| \le o(\sqrt{n})$. Also, note that for $w \ge 0$, the left-hand side of (5.1.22) monotonically increases. Therefore, (5.1.22) implies that there exists a constant $C_1 > 0$ such that for $0 \le w \le C/\sqrt{n}$,

$$
\mathbf{P}_n^{z_0}\{nu_1 < S_n \le nu_1 + \sqrt{n}\cdot w\} \le C_1 n\cdot\psi(z_0)^{-1} z_0 \cdot C_a \sqrt{n}\cdot w\cdot(nu_1)^{-a}.
$$

In turn, this bound implies that

$$
\text{(5.1.23)} \quad \begin{aligned} &\exp\{-n\cdot H(u_1)\}\cdot\int_0^{C/\sqrt{n}} \mathbf{P}_n^{z_0}\{nu_1 < S_n \le nu_1 + \sqrt{n}\cdot w\} \\ &\cdot d\left(\exp\{-z_0\sqrt{n}w + n\cdot f(u_1 + w/\sqrt{n})\}\right) \\ &\le \exp\{-n\cdot\mathbf{H}(u_1)\}\cdot C_1\cdot n\cdot\psi(z_0)^{-1}\cdot z_0\cdot C_a\cdot\sqrt{n}\cdot(nu_1)^{-a} \\ &\cdot\int_0^{C/\sqrt{n}} \left|\frac{d}{dw}\exp\{-z_0\sqrt{n}w + n\cdot f(u_1 + w/\sqrt{n})\}\right|\cdot w\cdot dw \end{aligned}
$$

(by analogy with (5.1.21′))

$$
\begin{aligned} &\sim \exp\{-n\cdot\mathbf{H}(u_1)\} C_1\cdot n\cdot\psi(z_0)^{-1}\cdot z_0\cdot C_a\cdot\sqrt{n}\cdot(nu_1)^{-a} \\ &\cdot\int_0^{C/\sqrt{n}} \left(-z_0\cdot\sqrt{n} + n/\sqrt{n}\cdot\left(f'(u_1) + \theta\cdot|f''(u_1)|\cdot w/\sqrt{n}\right)\right)\cdot w \\ &\cdot\exp\{-z_0\cdot\sqrt{n}\cdot w + n\cdot f(u_1) + \sqrt{n}\cdot f'(u_1)\cdot w - n\cdot|f''(u_1)/2|\cdot w^2/n\}\cdot dw. \end{aligned}
$$

Recall that $f'(u_1) = z_0$. Hence, the rightmost expression in (5.1.23) does not exceed

$$
\text{(5.1.24)} \quad \begin{aligned} &\exp\{-n\cdot\mathbf{H}(u_1)\}\cdot Const\cdot n^{3/2-a}\cdot\exp\{n\cdot f(u_1)\}\cdot\int_0^{C/\sqrt{n}} w\cdot dw \\ &\le \exp\left\{n\cdot\left(f(u_1) - \mathbf{H}(u_1)\right)\right\}\cdot Const\cdot n^{3/2-a}/n. \end{aligned}
$$

It is clear that the expression on the right-hand side of (5.1.24) will be included in the remainder.

Following along the same lines, we can get an upper bound for the integral over the range $\sqrt{n}/\delta(n) \le w \le \sqrt{n}\cdot(u_3 - u_1)$. We derive that the corresponding integral is also included in the remainder.

Let us now evaluate the main integrals (from $-\sqrt{n}/\delta(n)$ to $-C_1/\sqrt{n}$ and from $C_2/\sqrt{n}$ to $\sqrt{n}/\delta(n)$). Applying formula (5.1.22) we obtain that the difference of these two integrals, multiplied by $\exp\{-n\cdot\mathbf{H}(u_1)\}$, is equivalent to

$$(5.1.25)\quad \begin{aligned} &-\exp\{-n\cdot\mathbf{H}(u_1)\}\cdot n\cdot\psi(z_0)^{-1}\cdot z_0\cdot C_a\cdot\sqrt{n}\cdot(nu_1)^{-a} \\ &\cdot\int_{-\sqrt{n}/\delta(n)}^{\sqrt{n}/\delta(n)} w\cdot d\left(\exp\{-z_0\sqrt{n}w + n\cdot f(u_1 + w/\sqrt{n})\}\right). \end{aligned}$$

Integrating the above integral by parts we establish that both expressions generated by the term outside the integral will be included in the remainder. Therefore, it is only left to find the asymptotics of the following emerged integral:

$$(5.1.26)\quad -\int_{-\sqrt{n}/\delta(n)}^{\sqrt{n}/\delta(n)} \exp\{-z_0\sqrt{n}w + n\cdot f(u_1 + w/\sqrt{n})\}\cdot dw.$$

But since $f'(u_1) = z_0$ and

$$n\cdot f(u_1 + w/\sqrt{n}) - n\cdot f(u_1) = \sqrt{n}f'(u_1)w + f''(u_1)\cdot w^2/2 + O(w^3/\sqrt{n})$$

as $n \to \infty$, then an application of the Laplace method yields that the expression (5.1.26) is equivalent to

$$\begin{aligned} &-\exp\{n\cdot f(u_1)\}\cdot\int_{-\sqrt{n}/\delta(n)}^{\sqrt{n}/\delta(n)} \exp\{-f''(u_1)\}\cdot w^2/2\cdot dw \\ &\sim -\exp\{n\cdot f(u_1)\}\cdot\sqrt{\frac{2\pi}{|f''(u_1)|}} \end{aligned}$$

as $n \to \infty$. Finally, a combination of this representation with (5.1.25) implies the assertion of the theorem. □

Remark 5.1.7. Note that in Theorems 5.1.1 and 5.1.3, we have excluded the case of

$a_1 = 3$ from our consideration. However, it is relatively easy to see that one can also cover this case by making the Cramér's transformation and using the results of Chapter 2 related to the asymptotics of the probabilities of large deviations in the case of power tails with index $\alpha_1 = a_1 - 1 = 2$.

5.2 Rough Asymptotics of the Probabilities of Large Deviations for I.I.D. Random Sums in the Case when Cramér's Condition is Fulfilled Only on a Finite Interval.

In this section, we obtain the rough asymptotics (up to logarithmic equivalence) of the probabilities of large deviations of S_n for fairly extensive classes of distributions $F(\cdot)$. Both the asymptotics, given by formula (0.17) of the Introduction, as well as the techniques used, go back to the famous papers by Cramér (1938) and Chernoff (1952). However, in contrast to these papers, we do not make assumptions regarding the upper limits of the values of large deviations. In addition, the results of this section will be used in the proof of Proposition 5.3.1 of the next section.

Now, in order to formulate our main result, let us introduce some notation. Recall that $V(z) = \log \psi(z)$, and let

$$z_0 := \sup\{z > 0: \psi(z) < \infty\}.$$

We say that Condition (A) is satisfied if

$$\text{(A)} \qquad V^{(4)}(z) \,/\, (V''(z))^2 \leq Const$$

in some left half-neighborhood of z_0, and that Condition (B) is satisfied if there exists $0 < \delta < 1$ such that

$$\text{(B)} \qquad V(z) \,/\, (z \cdot V'(z)) < 1 - \delta$$

in some left half-neighborhood of z_0.

The next result refines the theorem of Vinogradov (1985c).

Theorem 5.2.1. Assume that one of the following sets of conditions is satisfied:

i) $z_0 < \infty$, $\psi(z_0) < \infty$, and $\log \mathbf{P}\{X_i > x\} \sim -z_0 \cdot x$ as $x \to \infty$.

ii) $z_0 < \infty$, $\psi(z_0) = \infty$, and Conditions (A) and (B) are fulfilled.

Then for any real $\varepsilon > 0$,

(5.2.1) $\log \mathbf{P}\{S_n > y\} \sim -n\cdot\mathbf{H}(y/n)$

as $n \to \infty$ with $y \geq n\cdot(\mathbf{E}X_i + \varepsilon)$.

Proof of Theorem 5.2.1. Note that the upper estimate for both cases follows from the results of Chernoff (1952). Namely, we have that for all $y \geq n\cdot\mathbf{E}X_i$,

(5.2.2) $\mathbf{P}\{S_n > y\} \leq \exp\{-n\cdot\mathbf{H}(y/n)\}$.

Let us separately derive corresponding lower estimates for both cases.

i) The results by Bahadur and Rao (1960) cover the range of deviations $y \leq n\cdot(u_0 - \varepsilon)$ with an arbitrary (sufficiently small) positive ε. In order to cover the range of deviations $y > n\cdot(u_0 - \varepsilon)$, we need to consider two subcases:

i1) Let $u_0 < \infty$. Then it suffices to get a lower estimate only in the range of deviations $y \geq n\cdot u_0$ (for $y < n\cdot u_0$, such that $y/n \to u_0$, the same lower estimate would follow automatically).

It is clear that

(5.2.3) $\mathbf{P}\{S_n > y\} \geq n\cdot\mathbf{P}\{X_n > y - (n-1)\cdot u_0 + (n-1)\cdot\delta_1\} \cdot \mathbf{P}\{S_{n-1} > (n-1)\cdot(u_0-\delta_1)\}$

with a certain $\delta_1 > 0$. Recall that $\log \mathbf{P}\{X_i > x\} \sim -z_0\cdot x$ as $x \to \infty$ and that

$$\log \mathbf{P}\{S_{n-1} > (n-1)\cdot(u_0 - \delta_1)\} \sim -(n-1)\cdot\mathbf{H}(u_0 - \delta_1)$$

as $n \to \infty$, which is obtained by an application of the already quoted results by Bahadur and Rao (1960). This implies that the logarithm of the expression on the right-hand side of (5.2.3) is equivalent to

(5.2.4) $-(n-1)\cdot\mathbf{H}(u_0 - \delta_1) - z_0\cdot(y - (n-1)\cdot u_0 + (n-1)\cdot\delta_1)$.

Recall (see the Introduction) that for $u \geq u_0$,

$$\mathbf{H}(u) = \mathbf{H}(u_0) + z_0\cdot(u - u_0) = z_0\cdot u - V(z_0),$$

which in turn implies that expression (5.2.4) can be represented as

$$-n\cdot\mathbf{H}(y/n) - (n-1)\cdot(\mathbf{H}(u_0 - \delta_1) - \mathbf{H}(u_0) + z_0\cdot\delta_1) + \mathbf{H}(u_0) - z_0\cdot u_0.$$

Since the rate function $\mathbf{H}(\cdot)$ is continuous, for any real $\varepsilon > 0$, the following is valid:

$$|\mathbf{H}(u_0 - \delta_1) - \mathbf{H}(u_0) + z_0\cdot\delta_1| < \varepsilon/2,$$

where δ_1 is chosen from the condition $z_0\cdot\delta_1 < (\varepsilon/4) \wedge (u_0 - \mathbf{E}X_i)$. Thus, for any real $\varepsilon > 0$ and for all sufficiently large n, expression (5.2.4) is greater than or equal to

$$-n\cdot\mathbf{H}(y/n) - n\cdot\varepsilon.$$

A combination of this lower estimate with the upper estimate (5.2.2) implies the assertion of point i1).

i2) Let $u_0 = \infty$. Then the validity of representation (5.2.1) in the range of deviations $y/n \leq Const$ follows from the results of Bahadur and Rao (1960). Now, assume that $y/n \to \infty$. Then a combination of (0.7) and (0.8) implies that

(5.2.5) $$\log \mathbf{P}\{S_n > y\} \geq \log(n \cdot \mathbf{P}\{X_i > y\}/2) \geq -z_0 y + o(y).$$

Recall that $z_0 < \infty$, $u_0 = \infty$, and let $y = y_n$. We then have that

$$-n \cdot \mathbf{H}(y/n) = -n \cdot (z_n y_n/n - V(z_n)),$$

where z_n is the unique root to the equation $V'(z_n) = y_n/n$. Moreover, we know that $z_n \to z_0$, $n = o(y_n)$, and $V(z_n) = \log \psi(z_n) \leq \log \psi(z_0) < \infty$. Hence, one is certain that the expression $-n \cdot \mathbf{H}(y_n/n)$ can be rewritten as

(5.2.6) $$-n \cdot \mathbf{H}(y_n/n) = -z_0 y_n + (z_0 - z_n) y_n + n \cdot V(z_n).$$

It is clear, in view of (5.2.5), that it suffices to demonstrate that the expression on the right-hand side of (5.2.6) is equal to $-z_0 y_n + o(y_n)$ as $n \to \infty$. But the latter follows from the facts that $z_0 - z_n \to 0$ as $n \to \infty$ and that the sequence $V(z_n)$ monotonically increases to $V(z_0)$ ($< \infty$) as $z_n \to z_0$. This in turn implies that $V(z_n) = o(y_n/n)$. Hence, a combination of (5.2.5) and (5.2.6) yields that

$$\log \mathbf{P}\{S_n > y_n\} \geq -z_0 y_n + o(y_n) = -n \cdot \mathbf{H}(y_n/n) + o(y_n)$$

as $n \to \infty$, and for an arbitrary sequence $\{y_n\}$ such that $y_n/n \to \infty$. A combination of this lower estimate with the upper estimate (5.2.2) implies the assertion of point i2).

ii) It is clear that the fact that $z_0 < \infty$ and $\psi(z_0) = \infty$ implies that $V'(z) \uparrow \infty$ as $z \uparrow z_0$. Therefore, the part of the proof that pertains to the range of deviations $y/n \leq Const$ follows from the results of Bahadur and Rao (1960). Hence, it suffices to derive the lower estimate for an arbitrary subsequence $\{y_n\}$ such that $u_n := y_n/n \to \infty$ as $n \to \infty$. In addition, note that $u_n < u_0 = \infty$ (see the definition of u_0 above formula (0.18) in the Introduction). Hence, one is certain that

$$\mathbf{H}(u_n) = (y_n z_n)/n - V(z_n),$$

where z_n is the unique root to the equation $V'(z) = u_n$. An application of the Chebyshev inequality yields that for an arbitrary sequence $\{\delta_n\}$ of positive numbers,

$$(5.2.7)\qquad \mathbf{P}_n^{z_n}\{(S_n - y_n)/n > \delta_n\} \le \mathbf{V}_{z(n)}(X_i)/(n\cdot\delta_n^2) \le V''(z_n)/(n\cdot\delta_n^2).$$

Now note that

$$V'''(z_n) = \mathbf{E}_{z_n}(X_i - \mathbf{E}_{z_n}X_i)^3,$$

and

$$V^{(4)}(z_n) = \mathbf{E}_{z_n}(X_i - \mathbf{E}_{z_n}X_i)^4 - 3\cdot(\mathbf{V}_{z_n}X_i)^2.$$

Let us also keep in mind that Condition (**A**) is fulfilled. Then the following bound is derived by the use of the Berry-Essen and the Jensen inequalities:

$$|\mathbf{P}_n^{z_n}\{S_n > y_n\} - 1/2| \le C_1\cdot\left(V^{(4)}(z_n)/V''(z_n)^2 - 3\right)^{3/4}/\sqrt{n} \to 0$$

as $n \to \infty$, $y_n/n \to \infty$. In particular, the above relationships imply that

$$(5.2.8)\qquad \mathbf{P}_n^{z_n}\{S_n > y_n\} \ge 1/2 - C_2/\sqrt{n}$$

as $n \to \infty$, $y_n/n \to \infty$. Now, set

$$\delta_n := \sqrt{(V''(z_n)/\sqrt{n})}.$$

Then (5.2.7) implies that

$$(5.2.9)\qquad \mathbf{P}_n^{z_n}\{S_n > y_n + n\cdot\delta_n\} \le 1/\sqrt{n}.$$

A subsequent combination of (5.2.8) and (5.2.9) implies that for sufficiently large n,

$$\mathbf{P}_n^{z_n}\{y_n < S_n \le y_n + n\cdot\delta_n\} \ge 1/4.$$

Now, let us use (0.19′) along with the above inequality to obtain that

$$\mathbf{P}\{S_n > y\} \ge \psi(z_n)^n\cdot \int_{\{y_n < S_n \le y_n + n\cdot\delta_n\}} \exp\{-z_n\cdot x\}\cdot dF_{S_n}^{z_n}(x)$$

$$\ge \psi(z_n)^n\cdot\exp\{-z_n\cdot(y_n + n\cdot\delta_n)\}/4.$$

Taking logarithms of both sides and keeping in mind that $\mathbf{H}(y_n/n) = (y_n z_n)/n - \log\psi(z_n)$, we get that

$$(5.2.10)\qquad (1/n)\log\mathbf{P}\{S_n > y\} \ge -\mathbf{H}(y/n) - z_n\delta_n - (1/n)\log 4.$$

Therefore, in view of (5.2.10), it suffices to demonstrate that a combination of Conditions (**A**) and (**B**) implies that

$$(5.2.11)\qquad z_n\delta_n = o(\mathbf{H}(y/n))$$

as $n \to \infty$, where $\delta_n := (V''(z_n)/\sqrt{n})^{1/2}$.

Indeed, it follows from (**A**) that

(5.2.12) $\quad V''(z)/V'(z)^2 \leq Const$

in some left half-neighborhood of z_0. Moreover,

$$\frac{z_n \cdot \delta_n}{\mathbf{H}(y_n/n)} = n^{-1/4} \cdot \frac{V''(z_n)}{V'(z_n)^2 \cdot \left(1 - V(z_n)/(z_n \cdot V'(z_n))\right)^2}.$$

Now, applying (**B**) we obtain the result that there exists $0 < \theta < 1$ such that

$$\theta < (1 - V(z_n)/(z_n \cdot V'(z_n)))^2 < 1$$

for all sufficiently large n. Combining the latter inequality with (5.2.12) we ascertain that (5.2.11) is true, and the proof of point ii) is complete. □

Remark 5.2.1. Condition (**A**) is essential as the following example demonstrates, where this condition is not fulfilled. Namely, assume that the distribution F is absolutely continuous with density $p_{X_1}(\cdot)$ such that

$$p_{X_1}(x) = \begin{cases} Const \cdot x^{-1} \cdot \exp\{-z_0 \cdot x\} & \text{for } x > 1, \\ 0 & \text{otherwise}, \end{cases}$$

where the value of the constant is chosen in order to make the function $p_{X_1}(\cdot)$ a probability density.

On the other hand, Condition (**B**) is not excessively restrictive on the reason that the ratio $z \cdot V'(z)/V(z)$ is unbounded. The question whether this condition may be violated remains open.

5.3 Discontinuity of the Most Typical Paths of Random Step-Functions and the Generalized Concept of the Action Functional.

Recall that if Cramér's condition is fulfilled on the whole positive semi-axis then the main part of the probability of a large deviation, $\mathbf{P}\{S_n > y\}$, is generated by small and approximately equal individual summands. However, in this section we will show that under fulfilment of Condition (5.1.13) with $a > 3$, such interpretation is true only in the range of large deviations where the exact asymptotics of $\mathbf{P}\{S_n > y\}$ is given by the Cramér's formula (i.e., for $y \leq n \cdot u_0$). This can also be clarified by revealing the path properties of random step-functions. Recall (see the Introduction) that r.s.f. $\eta_n(t) := S_{[nt]}/n$, where $t \in [0,1]$. Then it turns out that for $u \in (\mathbf{E}X_i, u_0]$, the conditioned r.s.f. $(\eta_n(\cdot) \mid \eta_n(1) > u\}$ approaches the *linear non-random function* $f_u(t) := u \cdot t$ as $n \to \infty$. However, the corresponding result for $u > u_0$ is quite opposite. Namely, the maximal term $X_1(n)$ of the random sample $\{X_1, \dots , X_n\}$ is approximately equal to $n \cdot (u - u_0)$, whereas the rest of the sum is approximately equal to $n \cdot u_0$, where u_0 is the expectation of X_i with respect to the transformed measure $\mathbf{P}_1^{z_*}$. The results of the next Theorem 5.3.1ii and Corollary 5.3.1 make the above explanations rigorous. However, in order to formulate Theorem 5.3.1, we first need to introduce some notation.

Let

$$\theta_u(t) := u_0 \cdot t + (u - u_0) \cdot \mathbf{I}_{\{U \leq t\}},$$

where $\mathbf{I}_A$ is the indicator of the set A, and r.v. U is uniformly distributed in [0,1]. Recall (see the notation above formula (4.2.27)) that

$$\Delta_1 f := \max_{t \in [0,1]} \big(f(t) - f(t-)\big),$$

and that

$$\tau_1(f) := \min\,(t \in (0,1]: f(t) - f(t-) = \Delta_1 f).$$

Now, let us slightly modify r.s.f. $\eta_n(t)$ by eliminating its maximal jump, starting from the (random) time instant of the occurrence of this jump. Namely, for $t \in [0,1]$ consider r.s.f. $\eta_n^*(t)$ such that

$$\eta_n^*(t) := \begin{cases} \eta_n(t) & \text{if } t \in [0, \tau_1(\eta_n)); \\ \eta_n(t) - \Delta\eta_n^*(\tau_1(\eta_n)) & \text{if } t \in [\tau_1(\eta_n), 1]. \end{cases}$$

Theorem 5.3.1. Let $\{X_n, n \geq 1\}$ be non-negative i.i.d.r.v.'s, and Condition (5.1.13) be fulfilled with $a > 3$. Fix $u > \mathbf{E}X_i$. Then

i) If $u \leq u_0$ then for any real $\varepsilon > 0$,

(5.3.1) $$\mathbf{P}\Big\{ \sup_{t\in[0,1]} |\eta_n(t) - t\cdot u| \le \epsilon \mid \eta_n(1) > u \Big\} \to 1$$

as $n \to \infty$.

ii) If $u > u_0$ then for any real $\varepsilon > 0$,

(5.3.1′) $$\mathbf{P}\Big\{ \sup_{t\in[0,1]} |\eta_n^*(t) - t\cdot u_0| \le \epsilon \mid \eta_n(1) > u \Big\} \to 1$$

as $n \to \infty$.

Remark 5.3.1. Note that for $u > u_0$, the asymptotic behavior of the conditioned r.s.f. $\eta_n^*(\cdot)$ with respect to the transformed measure $\mathbf{P}_n^{z_*}$ is of the same character as that given by formula (0.15) of the Introduction.

Proof of Theorem 5.3.1.

i) It suffices to show that

(5.3.2) $$\mathbf{P}\Big\{ \sup_{t\in[0,1]} |\eta_n(t) - t\cdot u| > \epsilon, \eta_n(1) > u \Big\} = o\big(\mathbf{P}\{\eta_n(1) > u\}\big).$$

But the event $\{\eta_n(1) > u\} = \{S_n > n\cdot u\}$. Then an application of (5.1.12) yields that

(5.3.3) $$\mathbf{P}\{\eta_n(1) > u\} \sim C(u)\cdot\exp\{-n\cdot\mathbf{H}(u)\}\cdot n^{-1/2}$$

as $n \to \infty$, where $C(u)$ is a certain positive constant that depends on u. Note that $\eta_n(0) - 0\cdot u = 0$, i.e., the supremum presented under the probability sign on the left-hand side of (5.3.2) is not attained at 0. A combination of this fact with (5.3.3) implies that in order to establish (5.3.2) it suffices to show that

(5.3.4) $$\mathbf{P}\Big\{ \sup_{t\in[0,1]} |\eta_n(t) - t\cdot u| > \epsilon, \eta_n(1) > u \Big\} = o\big(\exp\{-n\cdot\mathbf{H}(u)\}\cdot n^{-1/2}\big)$$

as $n \to \infty$.

Now, it is clear that

$$\mathbf{P}\Big\{ \sup_{t\in(0,1]} |\eta_n(t) - t\cdot u| > \epsilon, \eta_n(1) > u \Big\}$$

$$\le \mathbf{P}\Big\{ \sup_{t\in(0,1]} (\eta_n(t) - t\cdot u) > \epsilon, \eta_n(1) > u \Big\}$$

$$(5.3.5)\qquad +\mathbf{P}\Big\{\sup_{t\in(0,1]}(t\cdot u-\eta_n(t))>\epsilon,\eta_n(1)>u\Big\}$$

$$\le \mathbf{P}\Big\{\max_{1\le k\le n}(S_k-k\cdot u)>n\cdot\epsilon,S_n>n\cdot u\Big\}$$

$$+\mathbf{P}\Big\{\max_{1\le k\le n}(k\cdot u-S_{k-1})>n\cdot\epsilon,S_n>n\cdot u\Big\}.$$

Note that the rightmost inequality in (5.3.5) is true on the reason that

$$\sup_{t\in(0,1]}(\eta_n(t)-t\cdot u)$$

is attained at a point ℓ/n, where $1\le\ell\le n$ is an integer, since r.s.f. $\eta_n(\cdot)$ is piecewise constant, and the function $-t\cdot u$ monotonically decreases. Similarly, one can show that

$$\sup_{t\in(0,1]}(t\cdot u-\eta_n(t)) = \max_{1\le k\le n}\big(k/n-\eta_n(k/n-)\big).$$

Let us separately estimate each of the probabilities emerged in the rightmost expression in (5.3.5). Applying the well-known formula for conditional probabilities (cf., e.g., Feller (1971) Vol. II) we get that

$$\mathbf{P}\Big\{\max_{1\le k\le n}(S_k-k\cdot u)>n\cdot\epsilon,S_n>n\cdot u\Big\}$$

$$(5.3.6)\qquad =\int_{nu}^{\infty}\mathbf{P}\Big\{\max_{1\le k\le n}(S_k-k\cdot u)>n\cdot\epsilon\mid S_n=x\Big\}\cdot p_{S_n}(x)\cdot dx.$$

Now, let us make the Cramér's transformation of measures by means of formula (0.19′′); this transformation is absolutely continuous on the σ-algebra generated by r.v.'s $(X_1, \dots, X_n)$. In addition, it is clear, in view of (0.19′), that

$$(5.3.7)\qquad p_{S_n}(x) = \psi(z(u))^n e^{-z(u)x}\cdot p_{S_n}^{z(u)}(x),$$

where $p_{S_n}^{z(u)}(\cdot)$ is the density of r.v. S_n with respect to the transformed measure $\mathbf{P}_n^{z(u)}$. Therefore,

$$\mathbf{P}\Big\{\max_{1\le k\le n}(S_k-k\cdot u)>n\cdot\epsilon\mid S_n=x\Big\}$$

$$(5.3.8)\qquad = \mathbf{P}_n^{z(u)}\Big\{\max_{1\le k\le n}(S_k - k\cdot u) > n\cdot\epsilon \mid S_n = x\Big\}.$$

A subsequent combination of (5.3.6) - (5.3.8) yields that

$$\mathbf{P}\Big\{\max_{1\le k\le n}(S_k - k\cdot u) > n\cdot\epsilon, S_n > n\cdot u\Big\}$$

$$(5.3.9)\qquad = \int_{nu}^{\infty} \psi(z(u))^n\cdot e^{-z(u)\cdot x}\cdot \mathbf{P}_n^{z(u)}\Big\{\max_{1\le k\le n}(S_k - k\cdot u) > n\cdot\epsilon \mid S_n = x\Big\}\cdot p_{S_n}^{z(u)}(x)\cdot dx$$

$$\le \psi(z(u))^n\cdot e^{-z(u)\cdot n\cdot u}\cdot \mathbf{P}_n^{z(u)}\Big\{\max_{1\le k\le n}(S_k - k\cdot u) > n\cdot\epsilon, S_n > n\cdot u\Big\}.$$

Recall that for $u \le u_0$, the rate function $\mathbf{H}(u) = u\cdot z(u) - V(z(u))$. Also, let us get rid of the condition $\{S_n > n\cdot u\}$ under the probability sign in the rightmost expression in (5.3.9). Recall that $\mathbf{E}_{z(u)}X_i = u$ and $\mathbf{V}_{z(u)}X_i = V''(z(u))$ (see the Introduction). Then an application of Lemma 1.1.4 implies that the rightmost expression in (5.3.9) does not exceed

$$(5.3.10)\qquad 2\cdot\exp\{-n\cdot\mathbf{H}(u)\}\cdot\mathbf{P}_n^{z(u)}\Big\{S_n - nu > n\cdot\epsilon - \sqrt{2\cdot n\cdot V''(z(u))}\Big\}.$$

The latter probability is easily estimated by the use of the Chebyshev inequality. Hence, we get that expression (5.3.10) and therefore the probability on the left-hand side of (5.3.6) do not exceed

$$(5.3.11)\qquad 2\cdot\exp\{-\mathbf{H}(u)\}\cdot\frac{n\cdot V''(z(u))}{\left(n\epsilon - \sqrt{2n\cdot V''(z(u))}\right)^2} = O\left(\exp\{-n\cdot\mathbf{H}(u)\}\cdot n^{-1}\right).$$

Applying a similar technique, one can derive the following bound for the second probability presented in the rightmost sum in (5.3.5):

$$\mathbf{P}\Big\{\max_{1\le k\le n}(k\cdot u - S_{k-1}) > n\cdot\epsilon, S_n > n\cdot u\Big\}$$

$$\le 2\exp\{-n\cdot\mathbf{H}(u)\}\cdot\mathbf{P}_n^{z(u)}\Big\{\max_{1\le k\le n}(k\cdot u - S_k) > n\cdot\epsilon - u\Big\}.$$

In turn, the latter probability can be easily estimated by the use of Lemma 1.1.4 with the subsequent application of the Chebyshev inequality. We have that

$$\mathbf{P}\Big\{\max_{1\le k\le n}(k\cdot u - S_k) > n\cdot\epsilon - u,\, S_n > n\cdot u\Big\}$$

(5.3.11′)

$$\le 2\cdot\exp\{-n\cdot\mathbf{H}(u)\}\cdot\frac{n\cdot V''(z(u))}{\Big(n\epsilon - u - \sqrt{2n\cdot V''(z(u))}\Big)^2} = \mathbf{O}\big(\exp\{-n\cdot\mathbf{H}(u)\}\cdot n^{-1}\big).$$

Combining (5.3.5), (5.3.6), (5.3.11) and (5.3.11′) we ascertain that relationship (5.3.4) is true. Recall that in order to complete the proof of point i) of the theorem it suffices to establish the validity of relationship (5.3.4). □

Proof of ii) involves the Cramér's transformation of measures with the subsequent application of Remark 5.1.6 to Theorem 5.1.2. In addition, we will use Theorem 2′ of Godovan'chuk (1981). In particular, this theorem will enable us to derive the exact asymptotics of the probabilities of large deviations of r.s.f. $\eta_n(\cdot)$ in $\mathbf{D}[0,1]$ (equipped with the uniform metric ρ) as $n\to\infty$, with respect to the transformed measure $\mathbf{P}_n^{z_0}$.

By analogy with (5.3.2), it suffices to show that

(5.3.12) $$\mathbf{P}\Big\{\sup_{t\in[0,1]}|\eta_n^*(t) - t\cdot u| > \epsilon,\, \eta_n(t) > u\Big\} = \mathbf{o}\big(\mathbf{P}\{\eta_n(1) > u\}\big).$$

Note that for $u > u_0$,

$$\mathbf{P}\{\eta_n(1) > u\} = \mathbf{P}\{S_n > n\cdot u\}$$

(5.3.13)

$$\sim n\cdot\mathbf{P}\{X_1 > n\cdot u - (n-1)\cdot u_0\}\cdot\exp\{-(n-1)\cdot\mathbf{H}(u_0)\} \sim C(u)\cdot\exp\{-n\cdot\mathbf{H}(u_0)\}\cdot n^{1-a}$$

as $n\to\infty$, where $C(u)$ is a certain positive constant that depends on u (this follows from Remark 5.1.6).

Now, the definition of r.s.f. $\eta_n^*(\cdot)$ implies that

$$\sup_{t\in[0,1]}|\eta_n^*(t) - t\cdot u_0|$$

(5.3.14)

$$= \Big(\sup_{t\in(0,\tau_1(\eta_n))}|\eta_n^*(t) - t\cdot u_0|\Big)\vee\Big(\sup_{t\in[\tau_1(\eta_n),1]}|\eta_n^*(t) - t\cdot u_0|\Big).$$

Here, by analogy with (5.3.4), we have used the fact that $\eta_n^*(0) - 0\cdot u = 0$, i.e., the supremum is not attained at 0. Now, let us make the Cramér's transformation of measures by means of formula (0.19′′) with $z = z_0$. Then the density $p_{S_n}^{z_0}(x)$ of r.v. S_n satisfies

(5.3.7) with $u = u_0$ ($< \infty$) and $z(u) = z_0$. Moreover, recall that the asymptotic behavior of $p_{S_n}(\cdot)$ is already revealed in Theorem 5.1.2. Also, note that the asymptotics of the right-hand tail of the density $p_{S_n}^{z_0}(x)$ of r.v. S_n with respect to $\mathbf{P}_n^{z_0}$ is of power type.

By analogy with the proof of point i), we should use the well-known formulas for conditional probabilities which were already used for the derivation of (5.3.6), (5.3.8), and (5.3.9). Subsequently, applying (5.3.14) we obtain that for any fixed real $\varepsilon > 0$,

$$\begin{aligned}
&\mathbf{P}\Big\{ \sup_{t\in[0,1]} |\eta_n^*(t) - t\cdot u_0| > \epsilon, \eta_n(1) > u \Big\} \le \exp\{-n\cdot \mathbf{H}(u)\} \\
(5.3.15)\quad &\cdot \mathbf{P}_n^{z_0}\Big\{ \Big(\sup_{t\in(0,\tau_1(\eta_n))} |\eta_n^*(t) - t\cdot u_0| \Big) \vee \Big(\sup_{t\in[\tau_1(\eta_n),1]} |\eta_n^*(t) - t\cdot u_0| \Big) > \epsilon, \\
&\eta_n(1) > u \Big\} \le \exp\{-n\cdot \mathbf{H}(u)\} \\
&\cdot \mathbf{P}_n^{z_0}\Big\{ \Big(\sup_{t\in(0,\tau_1(\eta_n))} |\eta_n^*(t) - t\cdot u_0| \Big) \vee \Big(\sup_{t\in[\tau_1(\eta_n),1]} |\eta_n^*(t) - t\cdot u_0| \Big) > \epsilon \Big\}.
\end{aligned}$$

Recall that r.v.'s $X_1, \ldots, X_n$ are non-negative and that the right-hand tail of their common distribution function $F^{z_0}(\cdot)$ with respect to the transformed measure $\mathbf{P}_n^{z_0}$ admits the following asymptotics as $x \to \infty$:

$$1 - F^{z_0}(x) \sim K\cdot x^{-(a-1)},$$

where K is a certain positive constant. In addition, note that the left-hand tail of d.f. $F^{z_0}(\cdot)$ decays exponentially. Let us point out that r.s.f. $\eta_n(\cdot)$ constructed starting from r.v.'s satisfying these properties was considered in Godovan'chuk (1981) (see Example 2 therein). Also, note that both the methods and the results of that work were already described in the Introduction and in Section 4.2. Now, let the set $\mathfrak{R} \subset \mathbf{D}[0,1]$ be defined as follows:

$$\begin{aligned}
\mathfrak{R} := \Big\{ &f\in D[0,1] : f(0) = 0, f \uparrow [0,1], \Big(\sup_{t\in(0,\tau_1(f))} |f(t) - t\cdot u_0| \Big) \\
&\vee \Big(\sup_{t\in[\tau_1(f)),1]} |f(t) - \Delta_1 f - t\cdot u_0| \Big) > \epsilon \Big\}.
\end{aligned}$$

Then it is relatively easy to verify that the conditions of Theorem 2′ of Godovan'chuk (1981) are satisfied with k = 2. In particular, it turns out that $\rho(\mathfrak{R},\mathbf{B}^1) > 0$. Therefore,

$$\mathbf{P}_n^{z_0}\Big\{\Big(\sup_{t\in(0,\tau_1(\eta_n))}|\eta_n^*(t)-t\cdot u_0|\Big)\vee\Big(\sup_{t\in[\tau_1(\eta_n)),1]}|\eta_n^*(t)-t\cdot u_0|\Big)>\epsilon\Big\}$$

$$=\mathbf{P}_n^{z_0}\{\eta_n^*(\cdot)\in\mathfrak{R}\} = O(n^2\cdot n^{-2\cdot(a-1)})$$

as $n\to\infty$. A subsequent combination of this relationship with (5.3.15) implies that

$$\mathbf{P}\Big\{\sup_{t\in[0,1]}|\eta_n^*(t)-t\cdot u_0|\Big)>\epsilon,\eta_n(1)>u\Big\}$$

$$\leq \exp\{-n\cdot\mathbf{H}(u)\}\cdot O(n^2\cdot n^{-2\cdot(a-1)})$$

as $n\to\infty$. In turn, a combination of the latter bound with (5.3.13) implies (5.3.12), since $a>3$. Recall that (5.3.12) is sufficient for the proof of point ii). □

It should be pointed out that Theorem 5.3.1ii yields the following

Corollary 5.3.1. Let all the conditions of Theorem 5.3.1 be fulfilled, and $u>u_0$. Then for any real $\varepsilon>0$,

$$\mathbf{P}\{\,|\Delta\eta_n^*(\tau_1(\eta_n))-(u-u_0)|\leq\varepsilon\mid\eta_n(1)>u\}\to 1$$

as $n\to\infty$.

Hence, if the event $\{\eta_n(1)>u\}$ occurs (with $u>u_0$) then a large deviation comes from one big jump approximately equal to $u-u_0$, whereas r.s.f. $\eta_n(\cdot)$ grows approximately as a linear function with slope of the tangent line equal to u_0 before and after the jump. Moreover, in view of the fact that r.v.'s $X_1,\ldots,X_n$ are independent and identically distributed, it is easily seen that the limiting distribution as $n\to\infty$ of the time instant $\tau_1(n)$ of the maximal jump of r.s.f. $\eta_n(\cdot)$ is in fact uniform on the interval [0,1].

It turns out that the above Corollary 5.3.1 is closely related to a generalization of the concept of the action functional introduced by Lynch and Sethuraman (1987). Let us explain this in details. Thus, it is well known (cf., e.g, Freidlin and Wentzell (1984) or Wentzell (1990)) that the rough asymptotics of $\mathbf{P}\{\eta_n(\cdot)\in A\}$ as $n\to\infty$ for certain sets $A\subset\mathbf{D}[0,1]$ (with $\rho(A,\mathbf{O})>0$) is expressed in terms of the classical action functional $\mathrm{I}(\cdot)$ (here, $\mathbf{O}$ is the function on [0,1] identically equal to 0). Namely,

$$(5.3.16) \qquad \log \mathbf{P}\{\eta_n(\cdot) \in A\} \sim -n \cdot \inf_{f \in A} \mathrm{I}(f)$$

as $n \to \infty$, where for our special case, the classical action functional $\mathrm{I}(\cdot)$ can be defined as follows:

$$\mathrm{I}(f) := \begin{cases} \int_0^1 \mathbf{H}(\dot{f}) \cdot dt & \text{for absolutely continuous and monotonically increasing } f \text{ such that } f(0) = 0; \\ +\infty & \text{otherwise.} \end{cases}$$

It is clear that (5.3.16) is a functional analog of (0.17).

Now, assume that all the conditions of Theorem 5.3.1 are fulfilled. In particular, this means that the trajectories of $\eta_n(\cdot)$ are monotonically increasing functions on $[0,1]$, equal to 0 if $t = 0$. Under these conditions, Theorem 5.1 of Lynch and Sethuraman (1987) enables one to generalize the above definition of $\mathrm{I}(\cdot)$ for a much wider class of functions. Namely, let

$$\hat{\mathrm{I}}(f) := \begin{cases} \int_0^1 \mathbf{H}(\dot{f}) \cdot dt + z_0\left(f(1) - f(0) - \int_0^1 \dot{f} \cdot dt\right) & \text{for monotonically increasing } f \text{ such that } f(0) = 0; \\ +\infty & \text{otherwise.} \end{cases}$$

In fact, Lynch and Sethuraman (1987) demonstrated that relationship (5.3.16) remains valid if $\mathrm{I}(\cdot)$ is replaced by $\hat{\mathrm{I}}(\cdot)$. Then, in view of (5.3.16) and this remark, the typical behavior of the trajectories of $\eta_n(\cdot)$ is determined by the extremum (or extrema) where $\hat{\mathrm{I}}(\cdot)$ attains its minimum. Set

$$\mathfrak{R}_u := \{ f \in \mathrm{D}[0,1] : f(1) > u \}.$$

Then the next Proposition demonstrates the principal difference between the cases of $u \le u_0$ and of $u > u_0$.

Proposition 5.3.1. Let all the conditions of Theorem 5.3.1 be fulfilled. Then

i) If $u \le u_0$ then

$$\inf_{f\in\Re_u} \mathring{\mathbf{I}}(f) = -\mathbf{H}(u) = -\big(z(u)\cdot u - V(z(u))\big),$$

and it is attained at the unique function $f_u(t) := u\cdot t$.

ii) If $u > u_0$ then

$$\inf_{f\in\Re_u} \mathring{\mathbf{I}}(f) = -\mathbf{H}(u) = -\big(z_0\cdot u - V(z_0)\big),$$

and it is attained as at the function $f_u(t)$ as well as at the functions having the first derivative equal to u_0 almost everywhere with finite (or countable) number of positive such that the sum of the jumps is equal to $u - u_0$.

Proof of Proposition 5.3.1 easily follows from (5.3.16) and Theorem 5.2.1i of the previous section, since the fulfilment of Condition (5.1.13) obviously implies the fulfilment of all the conditions of Theorem 5.2.1i. □

Remark 5.3.2. It should be mentioned that the result of Proposition 5.3.1 is closely related to the fact that under fulfilment of Condition (0.16), the rate function $\mathbf{H}(u)$ is not strongly downward convex (see the Introduction). In addition, note that Proposition 5.3.1 can be proved in a straightforward manner, viz., without application of the probabilistic representation (5.3.16) and Theorem 5.2.1i.

Remark 5.3.3. Note that in view of Proposition 5.3.1ii, for $u > u_0$ the contributions of the smooth function $f_u(t)$ and of the discontinuous functions having the first derivative equal to u_0 almost everywhere with finite (or countable) number of positive jumps whose sum is equal to $u - u_0$, are approximately the same if one is only interested in the rough (logarithmic) equivalence of the probabilities of large deviations. However, our Corollary 5.3.1 demonstrates that in fact only paths that are close to a path performing exactly one big jump of magnitude $u - u_0$ on the interval (0,1] and growing linearly with slope of the tangent line equal u_0 before and after the jump make essential contribution to the probability of event $\mathbf{P}\{\eta_n(\cdot) \in \Re_u\}$. The most typical paths of the conditioned r.s.f. $(\eta_n(\cdot) \mid \eta_n(\cdot) > 0)$ in both cases are shown in Fig. 7.

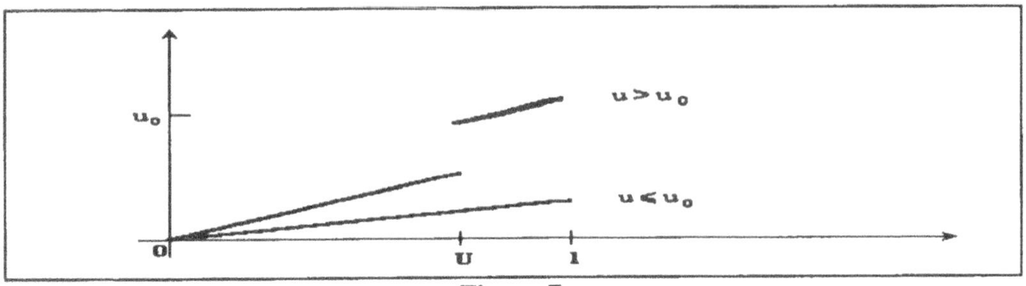

Figure 7

5.4 On Martingale Methods for the Derivation of Upper Estimates of the Probabilities of Large Deviations of I.I.D. Random Sums.

In this conclusive section, we use the martingale methods to derive upper bounds for the probabilities of large deviations of a certain family $\{M_k^N\}$ of reverse martingales (see formulas (5.4.14) - (5.4.15) below). It should be mentioned that subsequent terms of these random sequences are represented as infinite sums of i.i.d.r.v.'s. Note that the formulations of the results of this section were given in Dawson and Vinogradov (1994).

Now, let $\{E_n, n \geq 1\}$ denote a sequence of i.i.d.r.v.'s with common exponential (1) distribution. Set

$$B_n := \sum_{k=n+1}^{\infty} n \cdot E_k / (k \cdot (k-1)),$$

and

$$A_n^N := \sum_{k=n}^{N} (B_k - 1)^2 = \sum_{k=n}^{N} n^2 \cdot \left\{ \sum_{\ell=k+1}^{\infty} \frac{(E_\ell - 1)^2}{\ell^2 (\ell - 1)^2} \right\},$$

where $2 \leq n \leq N$. Recall that although these random sequences play only an auxiliary role in this section, their consideration was motivated by the rapidly developing theory of measure-valued processes (cf. Dawson and Vinogradov (1994) for details). In the next result, we exhibit the asymptotic behavior of $\{B_n\}$ and $\{A_n^N\}$.

Proposition 5.4.1 (see Dawson and Vinogradov (1994) Lemma 5.1). For any integer $n \geq 1$, r.v. $B_n < \infty$, **P**-*a.s.*, and

i) For any real $0 < \nu < 1/2$, there exists $C = C(\nu) > 0$ such that for any integer $n \geq 1$,

$$\mathbf{P}\{|B_n - 1| > n^{-\nu}\} \leq 2\cdot\exp\{-3n^{1-2\nu}/4\} + C/n^{1+\nu}.$$

ii) For any $0 < \nu < 1/2$, there exist $C = C(\nu)$ and an integer n_0 ($= n_0(\nu)$) such that for any integer $n \geq n_0(\nu)$,

$$|B_n - 1| \leq C\cdot n^{-\nu}, \quad \mathbf{P}\text{-}a.s.$$

iii) $$A_2^N/((\log N)/3) \to 1$$

as $N \to \infty$, $\mathbf{P}$-*a.s.*, which in turn implies that for any **fixed** integer $k \geq 2$, the random sequence A_k^N diverges towards $+\infty$ as $N \to \infty$, $\mathbf{P}$-*a.s.*

Proof of Proposition 5.4.1i is straightforward and is split into 3 steps. In steps 1 and 2, we establish an upper bound for the probabilities of right-hand large deviations of B_n from 1, while in step 3 an upper bound for the probabilities of left-hand large deviations s derived. Note that $\mathbf{E}B_n = 1$ and $\mathbf{V}(B_n) = O(1/n)$ as $n \to \infty$. Then, by definition,

$$\mathbf{P}\{B_n - 1 > n^{-\nu}\} = \mathbf{P}\left\{n\cdot\sum_{k=n+1}^{[n^{1+\nu}]}\frac{E_k}{k(k-1)} + n\cdot\sum_{k=[n^{1+\nu}]+1}^{\infty}\frac{E_k}{k(k-1)} > 1 + n^{-\nu}\right\}.$$

Step 1. Set

$$\zeta_n := n\cdot\sum_{[n^{1+\nu}]+1}^{\infty}\frac{E_k}{k(k-1)}.$$

Then $\mathrm{E}\zeta = n/[n^{1+\nu}]$. We will now show that

(5.4.1) $$\mathbf{P}\{(\zeta_n - \mathrm{E}\zeta_n) > n^{-\nu}\} \leq C/n^{1+\nu}.$$

In fact, the probability on the left-hand side of (5.4.1) is equal to

$$\mathbf{P}\left\{n\cdot\left\{\sum_{k=[n^{1+\nu}]+1}^{\infty}\left(\frac{E_k}{k(k-1)} - \frac{1}{[n^{1+\nu}]}\right)\right\} > n^{-\nu}\right\}$$

$$= \mathbf{P}\left\{\frac{n}{[n^{1+\nu}]}\cdot[n^{1+\nu}]\cdot\sum_{k=[n^{1+\nu}]+1}^{\infty}\frac{(E_k-1)}{k(k-1)} > n^{-\nu}\right\} \leq \frac{C\cdot n^{2\nu}\cdot n^2}{n^{1+\nu}(n^{1+\nu})^2} = C/n^{1+\nu}$$

by the Chebyshev inequality. □

Step 2. Recall that for $z < 1/\theta$,

$$\mathbf{E}\exp\{z\theta E_k\} = 1/(1-\theta z).$$

Then for $z < n$,

$$\mathbf{E}\left\{\exp\left(z\cdot n\cdot\sum_{k=n+1}^{[n^{1+v}]}\frac{E_k}{k(k-1)}\right)\right\} = \prod_{k=n+1}^{[n^{1+v}]}\frac{1}{1-\dfrac{n\cdot z}{k(k-1)}} < \infty.$$

Now, consider

$$\mathbf{P}\left\{n\sum_{k=n+1}^{[n^{1+v}]}\frac{E_k}{k(k-1)} > 1 + n^{-v}\right\}$$

$$\text{(5.4.2)} \qquad \le \exp\left\{-zn^{-v} + \sum_{k=n+1}^{[n^{1+v}]}\left(-\log\left(1-\frac{nz}{k(k-1)}\right) - \frac{nz}{k(k-1)}\right)\right\}$$

for any $z < n$, by the exponential Chebyshev inequality. A subsequent application of Taylor's formula yields that the left-hand side of (5.4.2) does not exceed

$$\exp\left\{-zn^{-v} + \sum_{j=2}^{\infty}\frac{(nz)^j}{j^2 n^{2j-1}}\right\} \le \exp\left\{-zn^{-v} + \frac{n}{2}\cdot\sum_{j=2}^{\infty}(z/n)^j\right\}$$

(5.4.3)

$$\le \exp\{-z\cdot n^{-v} + z^2/n\}$$

(if $0 < z < n/2$)

$$\le \exp\{-(3/4)\cdot n^{1-2v}\}$$

(choose $z = n^{1-v}/2$). In turn, steps 1 and 2 imply that

$$\text{(5.4.4)} \qquad \mathbf{P}\{B_n > 1 + n^{-v}\} \le C/n^{1+v} + \exp\{-(3/4)n^{1-2v}\}.$$

Step 3. Recall that for each $z > 0$,

$$\mathbf{E}\exp\{-z\theta E_k\} = 1/(1+\theta z).$$

Now, consider

$$\mathbf{P}\{B_n < 1 - n^{-v}\} = \int_0^{1-n^{-v}} dF_{B_n}(y) = \int_0^{1-n^{-v}} e^{zy}\cdot e^{-zy}\cdot dF_{B_n}(y)$$

$$\le e^{z(1-n^{-v})}\cdot\int_0^{\infty} e^{-zy}\cdot dF_{B_n}(y)$$

(with $z > 0$)

$$= e^{z(1-n^{-v})} \cdot \exp\left\{\sum_{k=n+1}^{\infty}\left(-\log\left(1+\frac{nz}{k(k-1)}\right)\right)\right\}$$

$$= \exp\left\{-zn^{-v} + \sum_{k=n+1}^{\infty}\left(-\log\left(1+\frac{nz}{k(k-1)}\right)+\frac{nz}{k(k-1)}\right)\right\}$$

$$= \exp\left\{-zn^{-v} + \sum_{k=n+1}^{\infty}\sum_{j=2}^{\infty}\frac{(-1)^j}{j}\cdot\left(\frac{nz}{k(k-1)}\right)^j\right\}$$

$$= \exp\left\{-zn^{-v} + \sum_{j=2}^{\infty}\frac{(-1)^j(nz)^j}{j}\cdot\sum_{k=n+1}^{\infty}\left(\frac{1}{k(k-1)}\right)^j\right\}$$

$$\le \exp\left\{-zn^{-v} + \sum_{j=2}^{\infty}\frac{(-1)^j(nz)^j}{j}\cdot\frac{1}{n^{2j-1}(2j-1)}\right\}$$

$$\le \exp\left\{-zn^{-v} + n\cdot\sum_{j=2}^{\infty}\frac{z^j}{n^j\cdot j^2}\right\} \le \exp\left\{-zn^{-v} + \frac{n}{4}\cdot\frac{z^2/n^2}{1-z^2/n^2}\right\}$$

$$\le \exp\{-zn^{-v} + z^2/n\} \le \exp\{-(3/4)\cdot n^{1-2v}\}$$

(see (5.4.3)); the interchange of summation is justified by Fubini's theorem since all the double sums in the above inequalities are absolutely convergent. A combination of the latter estimate for $\mathbf{P}\{B_n < 1 - n^v\}$ with (5.4.4) yields the assertion of i). □

ii) is obtained by applying a Borel-Cantelli argument to the inequality of i). □

iii) involves simple transformations of the given series, a Borel-Cantelli argument, some upper estimates for moments and a modification of the Chebyshev inequality. Obviously,

$$\text{(5.4.5)} \qquad \sum_{n=2}^{N} n^2\cdot\left\{\sum_{k=n+1}^{\infty}\frac{(E_k-1)^2}{k^2(k-1)^2}\right\} = \sum_{k=3}^{\infty}\frac{(E_k-1)^2}{k^2(k-1)^2}\cdot\sum_{n=2}^{(k-1)\wedge N} n^2,$$

where the interchange of summation is justified by Fubini's theorem, since the summands are non-negative. Note that for any integer $k \ge 3$,

$$\int_3^{k\wedge(N+1)} x^2\cdot dx \ge \sum_{n=2}^{(k-1)\wedge N} n^2 \ge \int_2^{(k-1)\wedge N} x^2\cdot dx,$$

which implies that

$$\sum_{n=2}^{(k-1)\wedge N} n^2 = \frac{(k-1)^3 \wedge N^3}{3} + O\left(((k-1)\wedge N)^2\right)$$

as $(k\text{-}1) \wedge N \to \infty$. Hence, the series (5.4.5) can be represented as

$$\frac{1}{3}\cdot \sum_{k=3}^{\infty} (E_k - 1)^2 \cdot \frac{(k-1)^3 \wedge N^3}{k^2(k-1)^2} + \sum_{k=3}^{\infty} (E_k - 1)^2 \cdot O\left(\frac{(k-1)^2 \wedge N^2}{k^2(k-1)^2}\right)$$

(5.4.6)

$$= \frac{1}{3}\cdot\left\{\sum_{k=3}^{N} (1/k - 1/k^2)\cdot(E_k - 1)^2 + N^3 \cdot \sum_{k=N+1}^{\infty} \frac{(E_k - 1)^2}{k^2\cdot(k-1)^2}\right\}$$

$$+ \left\{\sum_{k=3}^{N} (E_k - 1)^2 \cdot O(1/k^2) + O(N^2)\cdot \sum_{k=N+1}^{\infty} \frac{(E_k - 1)^2}{k^2\cdot(k-1)^2}\right\}.$$

Obviously, the series $\sum_{k=2}^{\infty} \frac{(E_k - 1)^2}{k^2}$ is convergent by the two-series theorem. On the other hand, the sum

$$\sum_{k=2}^{N} \frac{(E_k - 1)^2}{k} = \sum_{k=2}^{N} \frac{E_k^2 - 2E_k}{k} + \sum_{k=2}^{N} \frac{1}{k} \sim \log N \tag{5.4.7}$$

$N \to \infty$, **P**-*a.s.*, since the leftmost sum on the right-hand side of (5.4.7) is comprised of the independent random variables having zero expectations and variances of order k^{-2}, and hence converges **P**-*a.s.* by the two-series theorem.

Now, it is relatively easy to see that

$$N^3 \cdot \sum_{k=N+1}^{\infty} \frac{(E_k - 1)^2}{k^2\cdot(k-1)^2} = N^3 \cdot \sum_{k=N+1}^{\infty} \frac{1}{k^2\cdot(k-1)^2} + N^3 \cdot \sum_{k=N+1}^{\infty} \frac{E_k^2 - 2\cdot E_k}{k^2\cdot(k-1)^2}.$$

It is clear that the first (non-random) series on the right-hand side of this equality converges to a finite limit as $N \to \infty$. Let us prove that the second series approaches zero as $N \to \infty$, **P**-*a.s.* To this end, we split the series

$$N^3 \cdot \sum_{k=N+1}^{\infty} \frac{E_k^2 - 2\cdot E_k}{k^2\cdot(k-1)^2}$$

into two parts. It is equal to

$$N^3 \cdot \sum_{k=N+1}^{N^2} \frac{E_k^2 - 2 \cdot E_k}{k^2 \cdot (k-1)^2} + N^3 \cdot \sum_{k=N^2+1}^{\infty} \frac{E_k^2 - 2 \cdot E_k}{k^2 \cdot (k-1)^2}.$$

The rightmost term converges to zero as $N \to \infty$, **P**-*a.s.* This is easily derived from the Chebyshev inequality with a subsequent application of the Borel-Cantelli arguments.

Now, in order to demonstrate that the leftmost term also converges to zero as $N \to \infty$, **P**-*a.s.*, we fix an arbitrary $\varepsilon > 0$ and estimate the following probability from above by the use of a modification of the Chebyshev inequality. We have

$$\mathbf{P}\left\{\left|N^3 \cdot \sum_{k=N+1}^{N^2} \frac{E_k^2 - 2 \cdot E_k}{k^2 \cdot (k-1)^2}\right| > \epsilon\right\} \le \frac{N^9}{\epsilon^3} \cdot \mathbf{E}\left|\sum_{k=N+1}^{N^2} \frac{E_k^2 - 2 \cdot E_k}{k^2 \cdot (k-1)^2}\right|^3$$

(5.4.8)

$$\le C \cdot \frac{N^9}{\epsilon^3} \cdot \left\{\left(\sum_{k=N+1}^{N^2} \mathbf{V}\left(\frac{E_k^2 - 2 \cdot E_k}{k^2 \cdot (k-1)^2}\right)\right)^{3/2} + \sum_{k=N+1}^{N^2} \mathbf{E}\left|\frac{E_k^2 - 2 \cdot E_k}{k^2 \cdot (k-1)^2}\right|^3\right\},$$

by Theorem 3 of Rosenthal (1970). Making some relatively simple computations we conclude that the expression on the right-hand side of (5.4.8) is of the order $N^{-3/2}$, i.e., it is the general term of a convergent series. A subsequent application of the Borel-Cantelli arguments yields that the expression

$$N^3 \cdot \sum_{k=N+1}^{N^2} \frac{E_k^2 - 2 \cdot E_k}{k^2 \cdot (k-1)^2}$$

also converges to zero as $N \to \infty$, **P**-*a.s.* A subsequent application of the same technique yields that the expression

$$\sum_{k=3}^{N} (E_k - 1)^2 \cdot O(1/k^2) + O(N^2) \cdot \sum_{k=N+1}^{N^2} \frac{(E_k - 1)^2}{k^2 \cdot (k-1)^2}$$

is bounded from above as $N \to \infty$, **P**-*a.s.* A combination of these arguments with (5.4.6) yields the assertion of iii). □

Now, let us introduce the following family of random sequences depending on parameter N:

$$M_2^N := \sum_{m=2}^{N} (E_m - 1)\cdot(B_m - 1) = \sum_{m=2}^{N} (E_m - 1)\cdot\left(m \cdot \sum_{k=m+1}^{\infty} \frac{E_k - 1}{k(k-1)}\right), \ldots,$$

$$M_n^N := \sum_{m=n}^{N} (E_m - 1)\cdot(B_m - 1),$$

where $2 \le n \le N$, and $M_{N+1}^N := 0$. It can be shown that for any fixed integer $N \ge 2$, the random sequence

$$\{M_{N+1}^N, M_N^N, \ldots, M_3^N, M_2^N\}$$

constitutes a square integrable martingale. In addition, the corresponding quadratic characteristics are equal to

$$A_n^N = \sum_{m=n}^{N} (B_m - 1)^2,$$

and $A_{N+1}^N := 0$ (this can be readily checked by recursion).

The proof of the next result, Theorem 5.4.1, on almost-sure convergence relies on the derivation of upper estimates for the probabilities of large deviations of the random sequences M_k^N. In the proof, we also employ the techniques of the stochastic exponentials $Z_k^N(\theta)$, which play an essential role in constructing estimates for the probabilities of large deviations of martingales and semimartingales (cf., e.g., formula (13.8) of Chapter 4 of Liptser and Shiryayev (1989), Liptser and Shiryayev (1990), or Liptser and Pukhalskii (1992)). In this respect, let us emphasize the fact that both stochastic exponentials considered in this section are defined only on finite intervals (cf. (5.4.11) and (5.4.16) below). In addition, following Shiryayev (1984), we refer to our Theorem 5.4.1 as a distinctive version of the strong law of large numbers for our family of reverse martingales $\{M_k^N\}$.

Theorem 5.4.1 (see Dawson and Vinogradov (1994) Lemma 5.3). The ratio

$$M_2^N/A_2^N \to 0$$

as $N \to \infty$, **P**-*a.s.*

Proof of Theorem 5.4.1 involves Proposition 5.4.1, Borel-Cantelli arguments, and the derivation of upper bounds for the probabilities of large deviations for our family of reverse martingales $\{M_k^N\}$. Recall that our method of the derivation of upper bounds is similar to that employed in Chapter 4, Section 13 of Liptser and Shiryayev (1989)). On the other hand, Theorem 4 in Chapter 7, Section 5 of Shiryayev (1984) establishes the almost-sure convergence of a square integrable martingale (but not a reverse martingale) normalized by its quadratic characteristic, to zero, provided the quadratic characteristic diverges. In that work, a different, namely, martingale convergence technique is used. However, it seems that the martingale convergence technique cannot be easily adapted to our reverse martingale setting.

Now, recall that the quadratic characteristics A_n^N of the square integrable martingale $\{M_N^N, \dots, M_3^N, M_2^N\}$ are equal to

$$A_n^N = \sum_{m=n}^{N} (B_m - 1)^2.$$

By Proposition 5.4.1iii, the random sequence A_k^N diverges **P**-*a.s.* as $N \to \infty$ for any fixed $k \geq 2$.

Obviously, in order to prove the assertion of the theorem, it suffices to establish that for some **fixed** k,

$$M_k^N / A_2^N \to 0$$

as $N \to \infty$, **P**-*a.s.* To this end, let us estimate the following probability (for any fixed positive ε):

(5.4.9) $$\mathbf{P}\{M_k^N / A_2^N > \varepsilon\}.$$

Note that Proposition 5.4.1iii yields that there exists an integer N_0 (= $N_0(k)$) such that for any integer $N \geq N_0(k)$,

(5.4.10) $$A_k^N > 1/4 \cdot \log N,$$

P-*a.s.* Set $\lambda := 2^{3/2}$.

Now, following Chapter 4, Section 13 of Liptser and Shiryayev (1989), consider the stochastic exponential Z_k^N corresponding to our reverse martingale $\{M_k^N\}$:

$$Z_k^N(\theta) := \prod_{\ell=k}^{N} \mathbf{E}\left\{\exp\{\theta\cdot(M_\ell^N - M_{\ell+1}^N) \mid \sigma(E_{\ell+1}, E_{\ell+2}, \dots)\right\}$$

(5.4.11)

$$= \prod_{\ell=k}^{N} \frac{\exp\{-\theta\cdot(B_\ell - 1)\}}{1-\theta\cdot(B_\ell - 1)}.$$

Note that the product on the right-hand side of (5.4.11) is finite only for

$$\theta < \min_{\substack{k\le \ell\le N:\\ B_\ell>1}} \frac{1}{B_\ell - 1}$$

(in the sequel, we will choose $\theta := \lambda = 2^{3/2}$).

Applying Proposition 5.4.1ii, we ascertain that we can choose k from the condition

$$\lambda = 2^{3/2} \le \min_{\substack{k\le \ell\le N:\\ B_\ell>1}} \frac{1}{B_\ell - 1},$$

and $k := 2$ if for any integer $\ell \ge 2$, r.v.'s $B_\ell \le 1$. Hence, for any $k \le \ell \le N$, the stochastic exponential $Z_\ell^N(\lambda)$ is finite, **P**-*a.s.* Therefore, the probability (5.4.9) can be rewritten as follows:

$$\mathbf{P}\{M_k^N/A_2^N > \epsilon\} = \mathbf{P}\{\exp\{\lambda\cdot M_k^N - \log Z_k^N(\lambda)\} > \exp\{\lambda\cdot\epsilon\cdot A_2^N - \log Z_k^N(\lambda)\}\}$$

(5.4.12)

$$= \mathbf{P}\left\{\exp\{\lambda\cdot M_k^N - \log Z_k^N(\lambda)\} > \exp\left(\lambda\cdot\epsilon\cdot A_2^N + \frac{\lambda^2}{2}\cdot A_k^N + \lambda^3\cdot O\left(\sum_{\ell=k}^{N}(B_\ell - 1)^3\right)\right)\right\}$$

by (5.4.11). Note that by Proposition 5.4.1ii, the random series

$$\sum_{\ell=k}^{N}(B_\ell - 1)^3$$

converges to a finite limit, **P**-*a.s.* In turn, this implies that the expression

$$\lambda^3\cdot O\left(\sum_{\ell=k}^{N}(B_\ell - 1)^3\right)$$

is bounded both from above and from below by certain finite positive constants, **P**-*a.s.*

Now, let us split the probability on the right-hand side of (5.4.12) into two parts:

$$(5.4.13)\quad \begin{aligned} &\mathbf{P}\Big\{\exp\{\lambda\cdot M_k^N-\log Z_k^N(\lambda)\}>\exp\Big(\lambda\cdot\epsilon\cdot A_2^N+\frac{\lambda^2}{2}\cdot A_k^N+\lambda^3\cdot O\Big(\sum_{t=k}^{N}(B_t-1)^3\Big)\Big)\Big\}\\ &+\mathbf{P}\Big\{\exp\{\lambda\cdot M_k^N-\log Z_k^N(\lambda)\}>\exp\Big(\lambda\cdot\epsilon\cdot A_2^N+\frac{\lambda^2}{2}\cdot A_k^N+\lambda^3\cdot O\Big(\sum_{t=k}^{N}(B_t-1)^3\Big)\Big)\\ &\cap\{A_k^N>\frac{1}{4}\cdot\log N\}\Big\}\\ &=\mathbf{P}\Big\{\exp\{\lambda\cdot M_k^N-\log Z_k^N(\lambda)\}>\exp\Big(\lambda\cdot\epsilon\cdot A_2^N+\frac{\lambda^2}{2}\cdot A_k^N+\lambda^3\cdot O\Big(\sum_{t=k}^{N}(B_t-1)^3\Big)\Big)\\ &\cap\{A_k^N\le\frac{1}{4}\cdot\log N\}\Big\}. \end{aligned}$$

Taking into account (5.4.10) we ascertain that for any integer $N \geq N_0(k)$, the rightmost probability on the right-hand side of (5.4.13) equals zero. The leftmost probability on the right-hand side of (5.4.13) in turn does not exceed

$$(5.4.14)\qquad \mathbf{P}\Big\{\exp\{\lambda\cdot M_k^N-\log Z_k^N(\lambda)\}>Const\cdot\exp\Big(\lambda\cdot\epsilon\frac{\log N}{4}+\frac{\lambda^2}{2}\cdot\frac{\log N}{4}\Big)\Big\}.$$

It can be easily verified that the ratios

$$\{1,\ \exp\{\lambda{\cdot}M_N^N\}/Z_N^N(\lambda),\ \dots\ ,\ \exp\{\lambda{\cdot}M_k^N\}/Z_k^N(\lambda)\}$$

form a martingale. Hence, by Doob's inequality for non-negative supermartingales, the probability (5.4.14) does not exceed

$$(5.4.15)\quad \begin{aligned} &\mathbf{E}\Big(\exp\{\lambda\cdot M_{N+1}^N-\log Z_{N+1}^N(\lambda)\}\Big)\cdot\Big\{C_1\cdot\exp\Big(\lambda\cdot\frac{\epsilon}{4}\cdot\log N+\frac{\lambda^2}{8}\cdot\log N\Big)\Big\}^{-1}\\ &\le C_2\cdot N^{-(\lambda^2/8+\epsilon\cdot\lambda/4)}. \end{aligned}$$

Keeping in mind that by our choice $\lambda = 2^{3/2}$, we obtain the result that for any positive ε, the power of N in the expression on the right-hand side of (5.4.15) is less than -1. Therefore, it is the general term of a convergent series. Hence, the probability (5.4.9) is also the general term of a convergent series. A subsequent application of the first Borel-Cantelli lemma implies that for any positive ε, at most a finite number of events $\{M_k^N/A_2^N > \varepsilon\}$ occurs, **P**-*a.s.* Thus, it only remains to estimate

$$\mathbf{P}\{M_k^N/A_2^N < -\varepsilon\} = \mathbf{P}\{-M_k^N/A_2^N > \varepsilon\}.$$

The study of this probability is similar to that of the probability (5.4.9). The required result is obtained by deriving upper estimates for the probabilities of right-hand large deviations for the reverse martingale $\{\hat{M}_n^N\}$, where

$$\hat{M}_n^N := -M_n^N = \sum_{m=n}^{N} (E_m - 1)\cdot(1 - B_m)$$

if $2 \le n \le N$,

$$\hat{M}_{N+1}^N := 0.$$

The quadratic characteristics of the martingale

$$\{\hat{M}_{N+1}^N, \hat{M}_N^N, \ldots, \hat{M}_3^N, \hat{M}_2^N\}$$

are also equal to A_n^N, and the corresponding stochastic exponential $\hat{Z}_k^N(\cdot)$ is given by

$$\hat{Z}_k^N(\theta) := \prod_{\ell=k}^{N} \frac{\exp\{-\theta\cdot(1-B_\ell)\}}{1-\theta\cdot(1-B_\ell)}, \tag{5.4.16}$$

which is finite for

$$\theta < \min_{\substack{k\le\ell\le N:\\ B_\ell>1}} \frac{1}{1-B_\ell}.$$

Now, by Proposition 5.4.1ii we can choose k from the condition

$$\lambda = 2^{3/2} \le \min_{\substack{k\le\ell\le N:\\ B_\ell>1}} \frac{1}{1-B_\ell},$$

and $k := 2$ if for any integer $\ell \ge 2$, r.v.'s $B_\ell \ge 1$, etc.

Hence, for any positive ε, there exists $N_1 = N_1(\varepsilon)$ such that for any $N \ge N_1(\varepsilon)$, we have that $\{\,|M_k^N/A_2^N| \le \varepsilon\}$, $\mathbf{P}$-*a.s.* □

References

- D.Z. Arov and A.A. Bobrov (1960). The extreme terms of a sample and their role in the sum of independent variables, *Theory Probab. Appl.* **5**, 377-396.
- R.R. Bahadur and R.R. Rao (1960). On deviations of the sample mean, *Ann. Math. Statist.* **31**, 1015-1027.
- B. von Bahr (1965). On the convergence of moments in the central limit theorem, *Ann. Math. Statist.* **36**, 808-818.
- V. Bentkus and M. Bloznelis (1989). Nonuniform estimate of the rate of convergence in the CLT with stable limit distribution, *Lithuanian Math. J.* **29**, 8-17.
- V. Bentkus and R. Zitikis (1990). Probabilities of large deviations for L-statistics, *Lithuanian Math. J.* **30**, 215-222.
- A. Bikjalis (1966). Estimates of the remainder term in the central limit theorem, *Lithuanian Math. J.* **6**, 323-346.
- N.H. Bingham, C.M. Goldie and J.T. Teugels (1987). Regular Variation, Cambridge University Press: Cambridge.
- K.G. Binmore and H.H. Stratton (1969). A note on characteristic functions, *Ann. Math. Statist.* **40**, 303-307.
- R.P. Boas (1967). Lipschitz behavior and integrability of characteristic functions, *Ann. Math. Statist.* **38**, 32-36.
- A.A. Borovkov and A.A. Mogulskii (1992, 1993). Large deviations and testing statistical hypotheses. I - IV, *Siberian Advances in Math.* Vol.2, No. 3 52-120, Vol. 2, No. 4 43-72, Vol. 3, No. 1 19-86, Vol. 3, No. 2 14-80.
- H. Chernoff (1952). A measure of asymptotic efficiency for test of a hypothesis based on the sum of observations, *Ann. Math. Statist.* **4**, 493-507.
- R.S. Chhikara and J.L. Folks (1989). The Inverse Gaussian Distribution. Theory, Methodology, and Applications. Marcel Dekker, Inc.: New York.
- V.P.Chistyakov (1964). A theorem on sums of independent positive random variables and its applications to branching random processes, *Theory Probab. Appl.* **9**, 640-648.
- J. Chover, P. Ney, and S. Wainger (1973). Degeneracy properties of subcritical branching processes. *Ann. Probab.* **1**, 663-673.
- J.P. Cohen (1982a). The penultimate form of approximation to normal extremes, *Adv. Appl. Prob.* **14**, 324-329.
- J.P. Cohen (1982b). Convergence rates for the ultimate and penultimate approximations in extreme-value theory, *Adv. Appl. Prob.* **14**, 833-854.
- H. Cramér (1938). Sur un nouveau théorème - limite de la théorie des probabilités, *Act. Sci. et Ind.* f. 736.
- H. Cramér (1963). On asymptotic expansions for sums of independent random variables with a limiting stable distribution, Sankhyā A25, 13-24.
- G. Christoph (1982). Asymptotic expansions in the case of a stable limit law. *II, Lithuanian Math. J.* **22**, 69-79.

- G. Christoph and W.Wolf (1992). Convergence Theorems with a Stable Limit Law. Mathematical Research vol.70, Akad. Verl.: Berlin.
- M. Csörgő, S. Csörgő, L. Horváth and D.M. Mason (1986). Normal and stable convergence of integral functionals of the empirical distribution function, *Ann. Probab.* **14**, 86-118.
- S. Csörgő, E. Haeusler and D.M. Mason (1991). The quantile - transform - empirical - process approach to trimming. In: Sums, Trimmed Sums and Extremes (M.G. Hahn, D.M. Mason and D.C. Weiner eds.). Birkhäuser: Boston, pp. 215-267.
- D.A. Dawson and V. Vinogradov (1994). Mutual singularity of genealogical structures of Fleming-Viot and continuous branching processes. In: The Dynkin Festschrift (M.I. Freidlin ed.). Birkhäuser: Boston.
- V.N. Dubrovskii (1976). Exact asymptotic formulas of Laplace type for Markov processes, *Soviet Math. Dokl.* **17**, 223-227.
- R. Durrett (1980). Conditioned limit theorems for random walk with negative drift, *Z. Wahr. verw. Geb.* **52**, 287-297.
- M.V. Fedoryuk (1987). Asymptotics: Integrals and Series. Nauka: Moscow.
- W. Feller (1971). An Introduction to Probability Theory and its Applications. Vol. I, II, 2d edition. Wiley: New York.
- R.A. Fisher and L.H.C. Tippett (1928). Limiting forms of the frequency distribution of the largest or smallest number of a sample, *Proc. Camb. Phil. Soc.* **24**, 180-190.
- K. Fleischmann and I. Kaj (1992). Large deviation probabilities for some rescaled superprocesses. Preprint No. 10, Inst. Angew. Anal. Stoch.: Berlin.
- M.I. Fortus (1957). A uniform limit theorem for distributions which are attracted to a stable law with index less than one, *Theory Probab. Appl.* **2**, 478-479.
- M.I. Freidlin and A.D. Wentzell (1984). Random Perturbations of Dynamical Systems, Springer: New York.
- D.Kh. Fuk (1973). Some probabilistic inequalities for martingales, *Siberian Math. J.* **14**, 131-137.
- M.U. Gafurov and I.M. Khamdamov (1987). An estimate of the rate of convergence in the law of large numbers for sums of order statistics and their applications. In: Proc. Ist World Congress of the Bernoulli Society (Yu. Prohorov and V. Sazonov eds.), Vol. II. VNU: Utrecht, pp. 489-492.
- J. Galambos (1978). The Asymptotic Theory of Extreme Order Statistics, Wiley: New York.
- J.L. Geluk (1992). Second order tail behaviour of a subordinated probability distribution, *Stoch. Proc. Appl.* **40**, 325-327.
- J.L. Geluk and A.G. Pakes (1991). Second order subexponential distributions, *J. Austral. Math. Soc. (Ser. A)* **51**, 73-87.
- B.V. Gnedenko (1943). Sur la distribution limite du terme maximum d'une série aléatoire, *Ann. Math.* **44**, 423-453.
- V.V. Godovan'chuk (1978). Probabilities of large deviations for sums of independent random variables attracted to a stable law, *Theory Probab. Appl.* **23**, 602-608.
- V.V. Godovan'chuk (1981). Asymptotic behavior of probabilities of large deviations

arising from the large jumps of a Markov process, *Theory Probab. Appl.* **26**, 314-327.

- V.V. Godovan'chuk and V. Vinogradov (1990). On large deviations for trimmed sums of independent random variables. In: Probab. Theory and Math. Statist. Proc. 5th Vilnius Conf. Vol. I (B. Grigelionis et al. eds.), VSP/Mokslas: Utrecht/Vilnius, pp. 424-432.
- E.I. Gumbel (1958). Statistics of Extremes, Columbia Univ. Press: New York.
- L. de Haan and A. Hordijk (1972). The rate of growth of sample maxima, *Ann. Math. Statist.* **43**, 1185-1196.
- P. Hall (1978). On the extreme terms of a sample from the domain of attraction of a stable law, *J. London Math. Soc.* **18**, 181-191.
- P. Hall (1979). On the rate of convergence of normal extremes, *J. Appl. Prob.* **16**, 433-439.
- P. Hall (1980). Estimating probabilities for normal extremes, *Adv. Appl. Prob.* **12**, 491-500.
- P. Hall (1982). Rates of Convergence in the Central Limit Theorem. Research Notes in Math. Vol. 62, Pitman: Boston.
- P. Hall (1983). Fast rates of convergence in the central limit theorem, *Z. Wahr. verw. Geb.* **62**, 491-507.
- H. Hatori, M. Maejima and T. Mori (1980). Supplement to the law of large numbers when extreme terms are excluded, *Z. Wahr. verw. Geb.* **52**, 229-234.
- C.C. Heyde (1968). On large deviation probabilities in the case of attraction to a non-normal stable law, Sankhyā A30, 253-258.
- T. Höglund (1970). On the convergence of convolutions of distributions with regularly varying tails, *Z. Wahr. verw. Geb.* **15**, 263-272.
- I.A. Ibragimov (1967). On Chebyshev-Cramér asymptotic expansions, *Theory Probab. Appl.* **12**, 455-469.
- I.A. Ibragimov and R.Z. Has'minskii, (1981). Statistical Estimation - Asymptotic Theory. Springer: New York.
- I.A. Ibragimov and Yu.V. Linnik (1971). Independent and Stationary Sequences of Random Variables, Wolters-Noordhoff: Groningen.
- P.G. Inzhevitov (1986). Nonuniform estimate of the remainder of an asymptotic expansion in a local theorem for densities in the case of a stable limit law, *Bull. Leningrad Univ. Ser. Math., Mech. and Astronom.* No **2**, 106-108.
- B. Jørgensen (1982). Statistical Properties of the Generalized Inverse Gaussian Distribution. Lecture Notes in Stat. Vol. 9, Springer: New York.
- L.V. Kim and A.V. Nagaev (1972). An estimate of the rate of convergence in a limit theorem, *Random Processes and Statistical Inferences* **2**, 73-76 (FAN: Tashkent).
- C. Klüppelberg (1989). Estimation of ruin probabilities by means of hazard rates. *Insurance: Mathematics and Economics* **8**, 279-285.
- R.Sh. Liptser and A.N. Shiryayev (1989). Theory of Martingales, Kluwer: Dordrecht.
- R.Sh. Liptser and A.A. Pukhalskii (1992). Limit theorems on large deviations for semimartingales, *Stochastics and Stoc. Rep.* **38**, 201-249.
- G.S. Lo (1989). A note on the asymptotic normality of sums of extreme values, *J.*

Statist. Planning Inference, **22**, 127-136.

- J. Lynch and J. Sethuraman (1987). Large deviations for processes with independent increments. *Ann. Probab.* **15**, 610-627.
- R.A. Maller (1982). Asymptotic normality of lightly trimmed sums - a converse, *Math. Proc. Cambridge Phil. Soc.* **18**, 535-545.
- R. Michel (1976). On the accuracy of nonuniform Gaussian approximation to the distribution functions of sums of independent and identically distributed random variables, *Z. Wahr. verw. Geb.* **35**, 337-347.
- A.A. Mogulskii (1993). Large deviations for processes with independent increments. *Ann. Probab.* **21**, 202-215.
- A.V. Nagaev (1963). Large deviations for a class of distributions, Limit Theorems of Probability Theory, pp. 56-68 (FAN: Tashkent).
- A.V. Nagaev (1969). Limit theorems taking into account large deviations when Cramér's condition fails, *Izv. Akad. Nauk UzSSR Ser. Fiz-Mat. Nauk 13* No. **6**, 17-22.
- A.V. Nagaev (1970). Probabilities of Large Deviations of Sums of Independent Random Variables. Doctor of Science Thesis, Tashkent.
- A.V. Nagaev (1970). The role of the extreme terms of the variation series in the formation of a large deviation of a sum of independent random variables, *Soviet Math. Dokl.* **11**, 972-974.
- A.V. Nagaev (1985). New theorems on large deviations under Cramér's condition. In: 4th Internat. Vilnius Conf. on Probab. Theory and Math. Statist. Abstract of Communications. Vol. II, pp. 234-235. Institute of Mathematics and Cybernetics: Vilnius.
- A.V. Nagaev (1989). New theorems on large deviations under fulfilment of Cramér's condition. Depon. VINITI No 6102-V89.
- S.V. Nagaev (1979). Large deviations of sums of independent random variables, *Ann. Probab.* **7**, 745-789.
- L.V. Osipov (1972). On asymptotic expansions of the distributions of sums of random variables with non-uniform bounds of the remainder term, *Bull. Leningrad Univ. Ser. Math., Mech. and Astronom.* No **1**, 51-59.
- V.V. Petrov (1965). On the probabilities of large deviations of sums of random variables, *Theory Probab. Appl.* **10**, 287-298.
- V.V. Petrov (1966). A generalization of Cramér's limit theorem, *Selected Translations on Math. Statistics and Probability* **6**, 1-8. AMS: Providence.
- V.V. Petrov (1975a). Sums of Independent Random Variables, Springer: Berlin.
- V.V. Petrov (1975b). A generalization of an inequality of Lévy, *Theory Probab. Appl.* **20**, 141-145.
- I.F. Pinelis (1981). A problem of large deviations in a space of trajectories, *Theory Probab. Appl.* **26**, 69-84.
- I.F. Pinelis (1985). On asymptotic equivalence of the probabilities of large deviations for sums and maximum of independent random variables, Limit Theorems of Probability Theory. *Proc. Inst. Math. Vol.* **5**, pp. 144-173 (Nauka: Novosibirsk).
- E.J.G. Pitman (1968). On the behaviour of the characteristic function of a probability

distribution in the neighbourhood of the origin, *J. Austral. Math. Soc. (Ser. A)* **8**, 423-443.

- A. Puhalskii (1991). On functional principle of large deviations. In: New Trends in Probab. and Statist. Vol. I (V. Sazonov and T. Shervashidze eds.), VSP/Mokslas: Utrecht/Vilnius, pp. 198-218.
- S.I. Resnick (1987). Extreme Values, Regular Variation, and Point Processes. Springer: Berlin.
- H.P. Rosenthal (1970). On the subspaces of L^p ($p > 2$) spanned by sequences of independent random variables, *Israel J. Math.* **8**, 273-303.
- L.V. Rozovskii (1993). Probabilities of large deviations on the whole axis, *Theory Probab. Appl.* **38**, 53-79.
- L. Saulis and V. Statulevičius (1991). Limit Theorems for Large Deviations, Kluwer: Dordrecht.
- A.N. Shiryayev (1984). Probability, Springer: New York.
- R.L. Smith (1982). Uniform rates of convergence in extreme-value theory, *Adv. Appl. Probab.* **14**, 600-622.
- J.L. Teugels (1981). Limit theorems on order statistics, *Ann. Probab.* **9**, 868-880.
- L.H.C. Tippett (1925). On the extreme individuals and the range of samples taken from a normal population, *Biometrika*, **17**, 364-387.
- S.G. Tkachuk (1973). Local limit theorems taking into account large deviations in the case of limit stable laws, *Izv. Akad. Nauk UzSSR Ser. Fiz-Mat. Nauk 17* No **2**, 30-33.
- S.G. Tkachuk (1975). A theorem on large deviations in the case of distributions with regularly varying tails, *Random Processes and Statistical Inferences* **5**, 164-174 (FAN: Tashkent).
- S.G. Tkachuk (1977). Limit Theorems for Sums of Independent Random Variables Belonging to the Domain of Attraction of a Stable Law, Candidate of Science Thesis, Tashkent.
- V. Vinogradov (1983). Probabilities of large deviations for sums of independent random variables in the case of exponential tails. In: Proc. 21st All-Union Scientific Students Conf. Math., pp. 14-18. Novosibirsk Univ.: Novosibirsk.
- V. Vinogradov (1984). Asymptotic expansions in limit theorems on large deviations. In: Proc. 22d All-Union Scientific Students Conf. Math., pp. 17-21. Novosibirsk Univ.: Novosibirsk.
- V. Vinogradov (1985a). Asymptotic expansions of the probability of large deviations for sums of independent random variables under violation of Cramér's condition, *Theory Probab. and Math. Statist.* **31**, 21-27.
- V. Vinogradov (1985b). Uniform and non-uniform estimates of the residual term in the limit theorems in the case of convergence to the stable laws. In: 4th Internat. Vilnius Conf. on Probab. Theory and Math. Statist. Abstract of Communications. Vol. IV, pp. 322-324. Institute of Mathematics and Cybernetics: Vilnius.
- V. Vinogradov (1985c). Crude asymptotic form of the probabilities of large deviations of sums of independent random variables under Cramér's condition. *Bull. Moscow Univ. Ser. I 40* No **3**, 87-91.

- V. Vinogradov (1990a). On logarithmic refining terms in limit theorems taking into account large deviations of sums of independent random variables. In: Probab. Theory and Math. Statist. Proc. 5th Vilnius Conf. Vol. II (B. Grigelionis et al. eds.), VSP/Mokslas: Utrecht/Vilnius, pp. 552-562.
- V. Vinogradov (1990b). Conditioned limit theorems on weak convergence and discontinuity of typical paths of random step-functions. In: 11th Prague Conf. Inf. Theory, Statist. Decision Functions and Random Proc. Abstract of Communications. Charles Univ.: Prague.
- V. Vinogradov (1992a). On asymptotic expansions in limit theorems on large deviations for sums of independent random variables in the case of power tails, *C.R. Math. Rep. Acad. Sci. Canada* **14**, 83-88.
- V. Vinogradov (1992b). Limit theorems for extreme order statistics: large deviations and asymptotic expansions with non-uniform estimates of remainders, *C.R. Math. Rep. Acad. Sci. Canada* **14**, 131-136.
- V. Vinogradov (1992c). A non-uniform estimate taking into account large deviations in the limit theorem on non-normal convergence to the normal law, *C.R. Math. Rep. Acad. Sci. Canada* **14**, 285-290.
- V. Vinogradov (1993). Large deviations for i.i.d. random sums when Cramér's condition is fulfilled only on a finite interval, *C.R. Math. Rep. Acad. Sci. Canada* **15**, 229-234.
- V. Vinogradov (1994). Large deviations for order statistics. In: Proc. Conf. Extreme Value Theory & Appl. Vol. III (J. Galambos ed.), NIST: Gaithersburg.
- V. Vinogradov and V.V. Godovan'chuk (1989). Large deviations of sums of independent random variables without several maximal summands, *Theory Probab. Appl.*, **34**, 512-515.
- A.D. Wentzell (Venttsel') (1985). Infinitesimal characteristics of Markov processes in a function space which describe the past, *Theory Probab. Appl.* **30**, 661-676.
- A.D. Wentzell (1990). Limit Theorems on Large Deviations for Markov Stochastic Processes. Kluwer: Dordrecht.
- R. Zitikis (1991). On large deviations for L-estimates. In: New Trends in Probab. and Statist. Vol. I (V. Sazonov and T. Shervashidze eds.), VSP/Mokslas: Utrecht/Vilnius, pp. 137-164.

Printed and bound by CPI Group (UK) Ltd, Croydon, CR0 4YY
23/10/2024
01778230-0018